William Benjamin Carpenter, David Francis Condie

On the Use and Abuse of Alcoholic Liquors

William Benjamin Carpenter, David Francis Condie

On the Use and Abuse of Alcoholic Liquors

ISBN/EAN: 9783743313590

Manufactured in Europe, USA, Canada, Australia, Japa

Cover: Foto ©Lupo / pixelio.de

Manufactured and distributed by brebook publishing software
(www.brebook.com)

William Benjamin Carpenter, David Francis Condie

On the Use and Abuse of Alcoholic Liquors

PRIZE ESSAY.

ON THE

USE AND ABUSE

OF

ALCOHOLIC LIQUORS.

IN HEALTH AND DISEASE.

By WILLIAM B. CARPENTER, M.D., F.R.S.,

EXAMINER IN PHYSIOLOGY IN THE UNIVERSITY OF LONDON, PROFESSOR OF MEDICAL
JURISPRUDENCE IN UNIVERSITY COLLEGE, AND AUTHOR OF
"THE PRINCIPLES OF PHYSIOLOGY," ETC. ETC.

WITH A PREFACE,

By D. F. CONDIE, M.D.,

SECRETARY OF THE COLLEGE OF PHYSICIANS OF PHILADELPHIA, AND AUTHOR OF A
"PRACTICAL TREATISE ON THE DISEASES OF CHILDREN," ETC. ETC.

PHILADELPHIA:
HENRY C. LEA.
1866.

PREFACE BY DR. CONDIE.

THE opinion that Alcoholic Liquors afford to the human system a
stimulus, which, if not absolutely necessary to its well-being, is.
nevertheless, beneficial, by promoting in the several organs a vigorous and healthful exercise of their respective functions, and by
enabling them, thus, to resist more effectually the various disturbing
agencies to which they are daily subjected, is one that has been long
entertained, and of the correctness of which a large portion of the
public still entertains a firm conviction. To the influence of this
opinion may be ascribed much of the intemperance that has prevailed
in the world, and it even now presents a formidable barrier to the
success of every effort at reform in respect to the use of intoxicating
liquors as a beverage.

All are agreed as to the baneful influence upon health and morals
resulting from the excessive use of alcoholic drinks, and of the importance of guarding against this abuse of them by every means
within our power. But so long as the opinion prevails, that in
moderate quantities the use of these drinks is both proper and
salutary, it will scarcely be possible to guard the masses against
indulgence in them to excess — every effort to stay the progress of
intemperance, with its attendant evils, disease, poverty, insanity
and crime, will be unavailing.

1 *

To test the truth of the opinion referred to, by an examination of the effects produced upon the human frame by the use of alcoholic drinks, whether in moderate or excessive doses, is the object of the present Essay. And we know of nothing that has been written upon this important question better calculated to eradicate the prejudices which still exist in respect to intoxicating liquors, and to prevent the habitual abuse of them, by showing that their occasional moderate use, so far from promoting the health and vigour of the human frame, or increasing its capacity to sustain bodily or mental labour, or to resist the extremes of cold and heat, and other depressing agencies, is, on the contrary, under all circumstances, rather injurious than beneficial.

The author of the Essay is one in every respect well qualified to accomplish satisfactorily the task he has undertaken, by his acknowledged familiarity, as a physiologist, with the functions of the human system in a state of health, and by his acquaintance, as an able and skilful practitioner of medicine, with the different agencies by which those functions are disturbed or impeded, and the normal condition of the living organism replaced by one of disease. He brings to the investigation all the light which science can shed upon it, as well as the accumulation of facts derived from experience and observation.

Although we may differ from him as to the value of alcoholic drinks as a remedy for the cure of disease, still, in all his leading conclusions, as to the effects of these liquors upon the corporeal, mental and moral functions of the healthy human system, he is fully borne out by the results of our own observations and experience, acquired during a long series of years' practice as a physician, and as an active participator in the effort at temperance reform, which, originating with a few philanthropists in the United States, speedily enlisted the co-operation of " thousands of the best and most talented individuals" of our own and other lands.

Believing that the promotion of Temperance may be effectually accomplished by enlightening men's understandings in regard to the actual effects of alcoholic drinks upon the body and the mind, and thus enlisting in its favour the strongest motives by which human actions are influenced — the promotion of happiness, the preservation of health and vigour of frame, and the prolongation of life — we recommend to all the present Essay of DR. CARPENTER, as one of the best text-books of Temperance extant. We are persuaded that its extensive circulation will do much towards bringing about the only result by which drunkenness can be banished from our midst — namely, entire abstinence from all alcoholic drinks.

By the publication of the present cheap edition, Messrs. Blanchard and Lea have placed the work within the reach of every one; and their effort thus to aid in the diffusion of sound temperance doctrines will, we trust, be seconded by the advocates of those doctrines throughout our country.

To popularize as far as possible this edition, the various technical phrases which occur in it have been explained, so as to render their meaning familiar to the unscientific reader.

D. F. C.

Philadelphia, October, 1853.

ADVERTISEMENT

ISSUED BY DIRECTION OF THE DONOR OF THE PRIZE.

A PRIZE OF ONE HUNDRED GUINEAS

WILL BE GIVEN FOR THE BEST ESSAY ON THE USE OF ALCOHOLIC LIQUORS IN HEALTH AND DISEASE.

The Essay must contain answers to the following questions :—

1st.—What are the effects, corporeal and mental, of Alcoholic Liquors on the healthy human system?

2nd.—Does physiology or experience teach us, that Alcoholic Liquors should form part of the ordinary sustenance of Man, particularly under circumstances of exposure to severe labour or to extremes of temperature? Or, on the other hand, is there reason for believing that such use of them is not sanctioned by the principles of science, or the results of practical observation?

3rd.—Are there any special modifications of the bodily or mental condition of Man, short of actual disease, in which the occasional or habitual use of Alcoholic Liquors may be necessary or beneficial?

4th.—Is the employment of Alcoholic Liquors necessary in the practice of Medicine? If so, in what diseases, or in what forms and stages of disease, is the use of them necessary or beneficial?

The Essay must be delivered to the undersigned address, on or before the 30th day of September, 1849.

Dr. John Forbes, F.R.S., Physician to the Queen's Household, Prince Albert, and the Duke of Cambridge; Dr. G. L. Roupell, F.R.S., Physician to St. Bartholomew's Hospital; and Dr. W. A. Guy, M.B., Cantab., Professor of Forensic Medicine, King's College, London, have kindly consented to act as Adjudicators.

Signed on behalf of the Donor,

CHARLES GILPIN,
THOMAS BEGGS.

6, *Bishopsgate Street Without,*
London, April, 1848.

ADJUDICATION.

FROM the fifteen MS. Essays on the Use and Abuse of Alcoholic Liquors, transmitted to us by Messrs. Beggs and Gilpin for adjudication, we have unanimously selected as the best, the one bearing the motto — *Mens sana in corpore sano.* We accordingly adjudicate to its author the Prize of One Hundred Guineas.

We also think it due to the Author of the Essay bearing the motto — *Quot homines tot sententiæ,* to record our opinion of its great merits.

We further deem it right to speak in terms of commendation of the Essay having five mottoes, the first of which is — *How use doth breed a habit in a man.*

JOHN FORBES, M.D.
(Signed)   G. L. ROUPELL, M.D.
WILLIAM A. GUY, M.B.

London, December 6th, 1849.

TABLE OF CONTENTS.

CHAPTER I.

2   (xiii)

CHAPTER II.

CHAPTER III.

CHAPTER IV.

PREFACE.

~~~~~~~~

THE circumstances under which the following Essay is given to the Public, are sufficiently explained by the preceding Advertisement and Adjudication; but the Author takes this opportunity of offering a short statement of the objects which he had especially in view in its composition.

The questions set forth in the Advertisement having been evidently drawn up with great care, and having been obviously intended to bring the whole subject of the ordinary as well as the medical employment of Alcoholic Liquors under discussion, the Author judged it advisable to follow the plan which they had marked out, by answering each of them *seriatim;* [1] although he was aware that, by so doing, a certain amount of repetition would be almost necessarily involved. He found, as he proceeded, that it would be impossible to maintain such a continuity in his argument, as would be desirable for its effectiveness; and he would therefore request his readers, *in limine,* [2] to keep the following issue in view, as those to which he is desirous of leading them.

In the *first* place, — That from a scientific examination of the *modus operandi* [3] of Alcohol upon the Human body, when taken in a *poisonous* dose, or to such an extent as to produce Intoxication we may fairly draw inferences with regard to the specific [4] effects

---

[1] *Seriatim* — in order.
[2] *In limine* — at the outset.
[3] *Modus operandi* — the mode or way in which it acts
[4] *Specific* — peculiar.

2 *   (xvii)

which it is likely to produce, when repeatedly taken in excess, but not to an immediately-fatal amount.

Secondly,—That the consequences of the *excessive* use of Alcoholic liquors, as proved by the experience of the Medical Profession, and universally admitted by medical writers, being precisely such as the study of its effects in poisonous and immediately-fatal doses would lead us to anticipate, we are further justified in expecting that the habitual use of smaller quantities of these liquors, if sufficiently prolonged, will ultimately be attended, in a large proportion of cases, with consequences prejudicial to the human system,—the morbid actions thus engendered being likely rather to be chronic,[1] than acute,[2] in their character.

Thirdly,—That as such morbid actions are actually found to be among the most common disorders of persons advanced in life, who have been in the habit of taking a "moderate" allowance of alcoholic liquors, there is very strong ground for regarding them as in great degree dependent upon the asserted cause; although the long postponement of their effects may render it impossible to *demonstrate* the existence of such a connexion.

Fourthly,—That the preceding conclusion is fully borne out by the proved results of the "moderate" use of Alcoholic liquors, in producing a marked liability to the acute forms of similar diseases in hot climates, where their action is accelerated by other conditions; and also by the analogous facts now universally admitted, in regard to the remotely-injurious effects of slight excess in diet, imperfect aeration of the blood,[3] insufficient repose, and other like violations of the Laws of Health, when habitually practised through a long period of time.

Fifthly,—That the capacity of the healthy Human system to sustain as much bodily or mental labour as it can be legitimately called upon to perform, and its power of resisting the extremes of Heat and Cold, as well as other depressing agencies, are not augmented by the use of Alcoholic liquors; but that, on the other

---

[1] *Chronic* — slow: of long continuance.

[2] *Acute* — rapid in progress.

[3] *Aeration of the blood* — the change produced in the blood by its being brought in contact with the atmospheric air during its passage through the lungs.

hand, their use, under such circumstances, tends positively to the impairment of that capacity.

Sixthly,—That where there is a deficiency of power, on the part of the system, to carry on its normal[1] actions with the energy and regularity which constitute health, such power can rarely be imparted by the habitual use of Alcoholic liquors; its deficiency being generally consequent upon some habitual departure from the laws of health, for which the use of Alcoholic liquors cannot compensate; and the employment of such liquors, although with the temporary effect of palliating the disorder, having not merely a remotely-injurious effect *per se,*[2] but also tending to mask the action of other morbific[3] causes, by rendering the system more tolerant of them.

Seventhly,—That, consequently, it is the duty of the Medical Practitioner to discourage as much as possible the *habitual* use of Alcoholic liquors, in however "moderate" a quantity, by all persons in ordinary health; and to seek to remedy those slight departures from health, which result from the "wear and tear" of active life, by the means which shall most directly remove or antagonize their causes, instead of by such as simply palliate their effects.

Eighthly,—That whilst the habitual use of Alcoholic liquors, even in the most "moderate" amount, is likely (except in a few rare instances) to be rather injurious than beneficial, great benefit may be derived in the treatment of Disease, from the *medicinal* use of Alcohol in appropriate cases; but that the same care should be employed in the discriminating selection of those cases, as would be taken by the conscientious practitioner in regard to the administration of any other powerful remedy which is poisonous in large doses.

The foregoing appear to the Author to be the conclusions legitimately deducible from the facts and arguments which he has brought forwards; it will be for his Professional readers to decide, how far the case which he has made out is sufficiently strong to lead them

---

[1] *Normal*—appropriate; proper; healthy.
[2] *Per se*—by or of themselves.
[3] *Morbific*—disease-producing.

to the same results. This much, however, he would add; that
when he first entered upon the investigation, some years ago,
he had adopted no foregone conclusion, and had, consequently, no
temptation to make the facts square with preconceived views; that
he has constantly endeavoured to treat the subject as one of purely
scientific inquiry, and has avoided mixing up any other considerations with those which presented themselves to him as a Physiologist
and a Physician; and that, for the sake of keeping himself free
from even the appearance of partizanship, he has never allied himself with any one of the Societies, which have been formed to carry
into practical effect the Total Abstinence principle, but has preferred to follow a perfectly independent course. He ventures to hope
that on these grounds he may claim some right to being candidly
heard, by those to whom this Essay is more especially addressed.

He cannot allow it to go forth, however, without expressing his
conviction, that, whilst there are adequate Medical reasons for Abstinence from the *habitual* use of even a "moderate" quantity of
Alcoholic liquors, there are also strong Moral grounds for Abstinence
from that *occasional* use of them, which is too frequently thought
to be requisite for social enjoyment, and to form an essential part
of the rites of hospitality. The experience of every Practitioner
must bring the terrible results of Intemperance frequently before
his eyes; but whilst he is thus rendered familiar with its consequences as regards *individuals*, few, save those who have expressly
enquired into the subject, have any idea of the extent of the *social*
evils resulting from it, or of the degree in which they press upon
every member of the community. The Author believes that he is
justified in the assertion, that among those who *have* thus enquired,
there is but one opinion as to the fact, that, of all the causes which
are at present conspiring to degrade the physical, moral, and intellectual condition of the mass of the people, there is not one to be
compared in potency with the *Abuse of Alcoholic liquors;* and
that, if this could be done away with, the removal of all other
causes would be immeasurably promoted. Every one who wishes
well to his kind, therefore, must be interested in the enquiry how
this monster-evil can be best eradicated.

Now the Author considers, that the best answer to this enquiry
has been found in the results of experience. A fair trial has been

given, both in this country and in the United States, to societies which advocated the principle of *Temperance,* and which enlisted in their support a large number of intelligent and influential men; but it has been found that little or no good has been effected by them, among the classes on whom it was most desirable that their influence should be exerted, except where those who were induced to join them really adopted the *Total Abstinence* principle. Though he agrees fully with those who maintain, that *if* all the world would be *really temperate,* there would be no need of Total Abstinence Societies, the Author cannot adopt the inference that those who desire to promote the Temperance cause may legitimately rest satisfied with this measure of advocacy. For sad experience has shown, that a large proportion of mankind *cannot,* partly for want of the self-restraint which proceeds from moral and religious culture, be temperate in the use of Alcoholic liquors; and that the reformation of those who have acquired habits of intemperance *cannot* be accomplished by any means short of entire Abstinence from fermented liquors. Further, experience has shown that, in the present dearth of effectual education among the masses, and with the existing temptations to Intemperance arising out of the force of example, the almost compulsory drinking-usages of numerous trades, and the encouragement which in various ways is given to the abuse of Alcoholic liquors, nothing short of Total Abstinence can prevent the continuance, in the rising generation, of the terrible evils which we have at present to deplore. And lastly, experience has also proved that this reformation cannot be carried to its required extent, without the co-operation of the educated classes; and that their influence can only be effectually exerted by *example.* There is no case in which the superiority of example over mere precept is more decided and obvious, that it is in this. "I practise total abstinence my self," is found to be worth a thousand exhortations; and the lamentable failure of the advocates who cannot employ this argument, should lead all those whose position calls upon them to exert their influence, to a serious consideration of the claims which their duty to society should set up, in opposition to their individual feelings of taste or comfort.

Among the most common objections brought against the advocates of the Total Abstinence principle, is the following, — " That

the abuse of a thing good in itself does not afford a valid argument against the right use of it." This objection has been so well
met by the late Archdeacon Jeffreys of Bombay, (in a letter to the
Bombay Courier,) that, as it is one peculiarly likely to occur to
the mind of his Medical readers, the Author thinks it desirable to
quote a part of his reply. — "The truth is," he says, "that the
adage is only true under certain general limitations; and that out
of these, so far from being true, it is utterly false, and a mischievous
fallacy.   And the limitations are these : — If it be found by experience, that, in the general practice of the times in which we live,
the abuse is only the solitary exception, whereas the right use is
the general rule, so that the whole amount of good resulting from
its right use exceeds the whole amount of evil resulting from its
partial abuse, then the article in question, whatever it be, is fully
entitled to the benefit of the adage ; and it would not be the absolute and imperative duty of the Christian to give it up on account
of its partial abuse.   This is precisely the position in which stand
all the gifts of Providence, and all the enjoyments of life ; for
there is not one of them which the wickedness of man does not
more or less abuse.   But, on the other hand, if it be found by experience that there is something so deceitful and ensnaring in the
article itself, or something so peculiarly untoward connected with
the use of it in the present age, that the whole amount of crime,
and misery, and wretchedness connected with the abuse of it greatly
exceeds the whole amount of benefit arising from the right use of
it ; then the argument becomes a mischievous fallacy, the article in
question is not entitled to the benefit of it, and it becomes the duty
of every good man to get rid of it."   After alluding to the evidence that this is pre-eminently the case with regard to Alcoholic
liquors, the Archdeacon continues, — "We have then established
our principle, in opposition to the philosophic adage ; taking the
duty of the citizen and the patriot, even on the lowest ground.
But Christian self-denial and Christian love and charity, go far beyond this.   St. Paul accounted one single soul so precious, that he
would on no account allow himself any indulgence that tended
to endanger a brother's soul.   'If meat make my brother to offend,
I will eat no meat while the world standeth, lest I make my brother

to. offend.' — ' It is good neither to eat flesh nor to drink wine, nor anything whereby thy brother stumbleth, or is offended, or is made weak.' And we must bear in mind that flesh and wine are here mentioned by Paul as ' good creatures of God;' they are not intended to designate things evil in themselves. This saying of St. Paul is the Charter of Teetotalism; and will remain the charter of our noble cause so long as the world endures, so long as there remains a single heart to love and revere this declaration of the holy self-denying Paul."

If, then, the Author should succeed in convincing his readers, that the "moderate" habitual use of Alcoholic liquors is not beneficial to the healthy Human system, — still more, if they should be led to agree with him that it is likely to be injurious, he trusts that they will feel called upon by the foregoing considerations, to advocate the principle of Total Abstinence, in whatever manner they may individually deem most likely to be effectual. He believes it to be in the power of the Clerical and Medical Professions combined, so to influence the opinion and practice of the educated classes, as to promote the spread of this principle among the "masses," to a degree which no other agency can effect. And he ventures to hope that, whether or not he carries his readers with him to the full extent of his own conclusions, he will at any rate have succeeded in convincing them that so much is to be said on his side of the question, that it can no longer be a matter of indifference what view is to be taken of it; and that as "universal experience" has been put decidedly in the wrong with regard to many of the supposed virtues of Alcohol, it is at any rate possible. that its other attributes rest on no better foundation. In his general view of the case, he has the satisfaction of finding himself supported by the recorded opinion of a large body of his Professional brethren; upwards of *two thousand* of whom, in all grades and degrees, — from the court physicians and leading metropolitan surgeons, who are conversant with the wants of the upper ranks of society, to the humble country practitioner, who is familiar with the requirements of the artizan in his workshop, and the labourer in the field, — have signed the following certificate :—

" We, the undersigned, are of opinion —

" 1. That a very large proportion of human misery, including poverty, disease, and crime, is induced by the use of Alcoholic or fermented liquors as beverages.

" 2. That the most perfect health is compatible with Total Abstinence from all such intoxicating beverages, whether in the form of ardent spirits, or as wine, beer, ale, porter, cider, &c., &c.

" 3. That persons accustomed to such drinks may, with perfect safety, discontinue them entirely, either at once, or gradually after a short time.

" 4. That Total and Universal Abstinence from Alcoholic beverages of all sorts would greatly contribute to the health, the prosperity, the morality, and the happiness of the human race."

No medical man, therefore, can any longer plead the *singularity* of the Total Abstinence creed, as an excuse for his non-recognition of it; and although a certain amount of moral courage may be needed for the advocacy and the practice of it, yet this is an attribute in which the Author cannot for a moment believe his brethren to be deficient. Judging from his own experience, indeed, he may say that he has found much less difficulty in the course he has taken, than he anticipated when he determined on it; and that he has met with a cordial recognition of its propriety, not merely on the part of those who participated in his opinions but did not feel called upon to act up to them in their individual cases, but also among others who dissented strongly from his scientific conclusions, and who consequently had no more sympathy with his principles than with his practice.

London, March, 1850.

Ὡς φρονίμοις λέγω, κρίνατε ὑμεῖς ὃ φημι.

ON ALCOHOLIC LIQUORS.

CHAPTER I.

WHAT ARE THE EFFECTS, CORPOREAL AND MENTAL, OF ALCO-HOLIC LIQUORS ON THE HEALTHY SYSTEM?

1 IN replying to this question, it will be desirable to proceed as systematically as possible; since the results of our inquiries upon the several points which it involves, will have to form the ground-work of our further investigations. We shall commence, therefore, by examining the influence of Alcohol upon the *physical, chemical,* and *vital* properties of the several components of the animal fabric; from a knowledge of which we shall derive important assistance in our appreciation of its effects upon the human system as a whole.

I. INFLUENCE OF ALCOHOL UPON THE PHYSICAL, CHEMICAL, AND VITAL PROPERTIES OF THE ANIMAL TISSUES AND FLUIDS.

2. The most important *physical* change which the contact of Alcohol effects in the softer animal tissues, is that of *corrugation ;*[1] which change is entirely due to the difference in the capillary attraction[2] of the tissue for alcohol and for water respectively. If animal membranes, a mass of flesh, or coagulated fibrine[3] be placed in alcohol in a fresh state, (in which they are thoroughly charged with water,) there are formed, at all points where water and alcohol meet, mixtures of the two; and as the animal texture absorbs much less of an alcoholic mixture than of pure water, a larger amount of water is of course expelled, than of alcohol taken up; and the first result

---

[1] *Corrugation*—a contracting or drawing together, so as to form wrinkles, puckering.

[2] *Capillary attraction*—the disposition of minute, hair-like tubes or orifices to draw fluids within them.

[3] *Fibrine*—that portion of the blood from which is formed the chief portion of the muscles of red-blooded animals; *Coagulated* — rendered solid, curdled, clotted.

3

is a shrinking of the animal substance. "Thus," says Professor Liebig, "9·17 grammes of bladder, fresh, that is, saturated with water, (in which are contained 6·95 grammes of water and 2·22 of dry substance,) when placed in forty cubic centimetres of alcohol, weigh at the end of twenty-four hours 4·73 grammes, and have consequently lost 4·44 grammes. For *one* volume of alcohol, therefore, retained by the bladder, rather more than *three* volumes of water have been expelled from it." [1]

3. This corrugating effect of alcohol will be usually increased by the coagulating influence which 'it will exert on whatever soluble albumen [2] the tissues may contain. Both these results will, of course, be proportioned in their degree to the state of concentration of the alcohol; but some such physical change must always take place in the walls of the stomach, whenever alcoholic fluids are introduced into it; and in the soft tissues of the body at large, wherever alcohol has found its way into the current of the circulation. And that such is actually the case, is proved by the experiments of Dr. Percy, [3] who found that when animals are poisoned by alcohol introduced into the stomach, the coats of that organ become so thoroughly imbued with it, throughout their whole thickness, that no washing or maceration can remove it. He found, also, that the tissues remote from the stomach become impregnated with alcohol, when it has passed into the current of the circulation; but on this point we shall dwell more at length hereafter.—(§ 17.)

4. The physical change just described must have an important influence upon the *chemical* relations of the tissues; since it is impossible that alcohol can be substituted, in however small a proportion, for their constituent water, without producing a decided alteration in their chemical properties, which must disturb the normal series of changes involved in their nutritive operations. Among the most important of the chemical changes which alcohol has the power of effecting, is the coagulation of soluble Albumen: and although it will rarely, if ever, be introduced into the mass of the blood, or into the serous fluids of the tissues, by any ordinary alcoholic potations, in a sufficiently concentrated state to effect this, yet we should anticipate that its presence, even in a very dilute form, *must* affect the chemical relations of albumen, and can scarcely do otherwise than retard that peculiar transformation by which it is converted into the more *vitalized* substance, Fibrine. That such is actually the case will be rendered probable by the considerations to be presently adduced.

5. No considerable changes of a physical or chemical nature can

---

[1] On the Motion of the Animal Juices, p. 11.

[2] *Albumen*—an animal substance similar to the white of egg.

[3] Experimental Inquiry concerning the Presence of Alcohol in the Ventricles of the Brain, p. 29.

take place in any of the animal tissues, without disordering their *vital* properties also; and we have now to inquire into the mode in which these properties are affected by the contact of Alcoholic liquids. In the first place, it would appear that the solidifiability[1] of the fibrine, which is its special vital endowment,[2] is impaired by the introduction of alcohol into the fluid which contains it; for when an animal has been killed by the injection of alcohol into the blood-vessels, the blood often remains fluid after death, or coagulates but imperfectly. (See the experiments of Dr. Percy, *op. cit.*) Now, as it is probable that nearly all the organized tissues are developed at the expense of the fibrine, it is obvious that anything which impairs its organizibility[3] must have an injurious influence upon the general nutritive operations; and we shall hereafter find confirmation of this inference in that peculiar condition of the system which results from excessive habitual indulgence in alcoholic potations, and of which the imperfect elaboration[4] of the fibrine is one of the special characteristics. (§ 63.)—But, secondly, we find that when alcoholic liquids are applied to living tissues, especially to the vascular surface of the skin or mucous membrane,[5] they induce redness, heat, and pain, indicating an increased determination of blood to the part. These symptoms vary in intensity, according to the state of concentration of the liquid, and the length of time during which it remains in contact with the surface; and they may pass on from this condition of *irritation* to one of actual *inflammation*.

6. Our best knowledge, however, of the influence of Alcohol upon the vital actions of the animal tissues, is derived from microscopic observations upon the circulation of blood in the web of the frog's foot. If alcohol be applied to this membrane in a *very dilute* state, its first effect is to quicken the movement of blood through the vessels, which are at the same time rather contracted than dilated. But this state of things gradually gives place to the opposite; for after a time, which varies with the degree of the dilution of the alcohol, the circulation becomes retarded, and the vessels dilated; and a further time elapses before the original condition is recovered. If the alcohol have been applied at first, however, in a less dilute form, the first stage is not observed; but a retardation of the flow of blood is immediately brought about, and a considerable dilation of the vessels

---

[1] *Solidifiability*—capacity to become solid.

[2] The *coagulation* of albumen and the *fibrillation* of fibrine are two entirely different processes; the former being a simply *physical* aggregation, the latter tending to produce an organized tissue, and being therefore of a *vital* nature.

[3] *Organizibility*—capacity to become converted into and form part of a living tissue.

[4] *Elaboration*—formation.

[5] *Mucous membrane*—the lining membrane of the nostrils, mouth and throat, stomach and intestines, &c.

takes place. The retardation may be such as to amount in some
parts to a complete stagnation; and here it is noticed that the red
corpuscles[1] become crowded together, and that their normal form is
lost; their colouring matter also being diffused through the liquor
sanguinis.[2] Around the parts in which the stagnation is witnessed
however, there is generally a border, in which the blood flows with
increased rapidity. Now this perverted state may gradually give
place to the normal condition,[3] if the stimulus be only applied for a
short time; the circulation being restored where it was deficient, and
the vessels gradually contracting to their proper size. But if the
contact of concentrated alcohol be prolonged, it becomes obvious that
the tissue has been killed; for the circulation is never re-established
in it, and it is at last separated by gangrene.[4] We rarely witness,
in cold-blooded animals, those consequences of the application of irri-
tants which properly constitute the *inflammatory* process; but this
process is liable to be excited in man, and in warm-blooded animals,
by the contact of alcoholic fluids with living tissues, if the contact be
sufficiently prolonged, and the alcohol sufficiently concentrated.

7. Now the inference to be drawn from the preceding details is
these;—that Alcohol, when applied to the living tissues in a sufficiently dilute form, *exalts* for a time their vital activity; but that
this exaltation is temporary only, and is followed by a corresponding
depression. And further, that when the alcohol is in a state of sufficient concentration to act more potentially,[5] its exhausting or depressing effect is manifested, without any previous stage of excitement. This inference we shall hereafter find to be in precise accordance with that to which we shall be conducted by observation of the
effects of alcohol upon the system at large; and we are justified,
therefore, in regarding alcohol as belonging to the class of *stimulants,* and as subject to the laws of their operation. It has been
affirmed by some that alcohol in small doses is *tonic;*[6] but of this
there is no adequate proof. The property of tonic remedies is to
increase the vital contractility of the animal solids in general, but
more especially that of the walls of the blood-vessels. Now although
some slight effect of this kind is at first manifested, after the application of very dilute alcohol to a living membrane, yet it is very transitory, and is succeeded by a much longer period of *diminution* of the
tonic contractility of the walls of the blood-vessels. And we shall

---

[1] *Red corpuscles*—that portion of the blood upon which its colour depends.
[2] *Liquor sanguinis*—the fluid portion of the blood in which the red corpuscles float.
[3] *Normal condition*—appropriate, usual, healthy condition.
[4] *Gangrene*—death and decomposition of the tissue; mortification.
[5] *Potentially*—powerfully.
[6] *Tonic*—that which increases permanently the activity of the functions
of life.

he eafter see that the supposed tonic properties of alcohol in small duses, (especially in the form of wine or malt liquor,) are really but a manifestation of its stimulant effects.

8. Various other experiments confirm this view of the effects of Alconol on the animal tissues; and those of Humboldt are particularly valuable, as regards its special capability of producing a temporary excitement of *nervous power.* " When the crural nerve," [1] he says, " of a full-grown and lively frog was immersed in alcohol, if the leg was already exhausted by galvanization, the alcohol evidently increased its excitability; and this increase was lasting (*i. e.* for a time), when it was quickly removed from the stimulating fluid. If the nerve was left in it for some time, its excitability was completely exhausted. Its application exhausted instantaneously the excitability of young animals — birds, worms, and insects. If the tail of an earth-worm or leech be dipped for only *four seconds* in alcohol, it becomes stiff and excitable as far as it is immersed; and although in frogs and puppies this state of rigidity could sometimes be removed, in these animals it never could." [2]

9. There are some peculiar effects of Alcohol upon the blood, besides its influence on the coagulability of the fibrine, of which it is proper that special mention should be made. When alcohol is mingled with fresh arterial blood, it darkens its colour, so as to give it more or less of the venous aspect.[3]—(§ 118.) And when this admixture is made under the microscope, it is perceived that the red corpuscles shrink, and that a considerable part of their contents becomes mingled with the liquor sanguinis. Now, although the peculiar functions of the red corpuscles have not yet been precisely determined by physiologists, there is no doubt whatever that they are among the most important constituents of the blood; and there is strong reason to believe that they are subservient on the one hand to the respiratory function, and on the other, either directly or indirectly, to the elaboration of the *plasma* or organizable material of the blood. It is highly improbable, then, that any considerable effect can be produced upon them, without seriously impairing the processes of aeration and nutrition; both of which, as we shall hereafter see, are prejudicially influenced in other ways, by the presence of alcohol in the blood.

Having thus considered the influence of Alcohol upon the properties and actions of the component tissues of the animal fabric, we shall inquire into its effects upon the living system as a whole.

---

[1] *Crural nerve*—nerve of the leg.
[2] Annals of Medicine, 1799, p. 265.
[3] *Venous aspect*—the blood in the arteries is of a bright vermilion; that in the veins, of a dark purple hue or aspect.

II. IMMEDIATE CONSEQUENCES OF THE EXCESSIVE USE OF ALCOHOLIC LIQUORS ON THE GENERAL SYSTEM.

Phenomena of Alcoholic Intoxication.

10. The term Intoxication is sometimes employed in this country to designate that series of phenomena which results from the action of all such poisons as first produce stimulation, and then narcotism : [1] of these, however, Alcohol is the type; and the term is commonly applied to alcoholic intoxication alone. It is worthy of notice, however, that the designation is now given by French writers to the series of remote or constitutional effects consequent upon the introduction of *any* poisonous agent into the .blood; thus we meet with the terms "arsenical intoxication," "iodine intoxication," and even "purulent intoxication." In fact, it is there considered an equivalent (as its etymology denotes) of our word *poisoning ;* and the fact that such a term should be in common use in this country, to designate the ordinary results of the ingestion [2] of alcoholic liquors, is not without its significance; for, if the classical term "intoxication" be habitually employed as the equivalent of the Saxon "drunkenness," we are justified in turning that classical term into English again, and in asserting that *the condition of drunkenness, in all its stages, is one of poisoning.* That such is indeed the case, will become obvious from an examination of its symptoms, and from a comparison of them with those of the cases in which a fatal result has supervened upon excessive indulgence in alcoholic liquors. To such an examination we shall now proceed; first, detailing the symptoms of the slighter forms of intoxication; then, those of the deeper; and, lastly, those of the severest cases; and afterwards inquiring into the pathological [3] state from which those symptoms proceed, and the *modus operandi* [4] of the agent that has produced it.

11. Among the *first* effects of the ingestion of Alcoholic liquors, in sufficient amount to produce their characteristic influence, are, in most persons, an increase in the force and rapidity of the heart's contractions; producing a full, frequent, and strong pulse. With this, there seems to be a general exaltation of the organic functions ; the appetite and the digestive power being increased, and the secretions augmented, especially those of the skin and kidneys. But it is obvious that the encephalic [5] centres of the nervous system are especially acted on by the stimulus, for we observe all the manifesta-

---

[1] *Narcotism*—stupor, insensibility.
[2] *Ingestion*—taking into the stomach.
[3] *Pathological*—diseased condition.
[4] *Modus operandi*—the manner in which it acts.
[5] *Encephalic*—within the head; the brain

tions of an excited action in them, such as talkativeness, rapidity and variety of thought, exhilaration of the spirits, animation of the features and gestures, flushed countenance, and suffusion of the eyes. During slight intoxication, the prevailing dispositions and pursuits are often made manifest; and hence the saying, " *In vino veritas.*[1] The irritable and ill-tempered become quarrelsome; the weak and silly are boisterous with laughter and mirth, and profuse in offers of service; and the sad and hypochondriacal readily burst into tears, and dwell on mournful topics. It sometimes happens, however, that men habitually melancholy become highly mirthful, when they have drunk enough to excite them; but this seems rather to be the case when the melancholy results from external depressing influences, than when it is constitutional; and hence it is that too many persons in circumstances of distress or difficulty have recourse to the bottle for temporary solace from their cares. If no more liquor be taken than is sufficient to produce this condition, it gradually subsides, and is followed by a state of the opposite character; the appetite, the digestive power, and the organic functions in general, being lowered in activity, the skin dry, the secretions diminished, the spirits depressed, and the power of mental exertion for a time impaired. For this condition, sleep, and abstinence (not merely from a renewal of the stimulus, but from more food than the stomach really demands), are the most effectual remedies.

12. The state of mental excitement just described, is very similar to the incipient stage of Phrenitis[2] or Mania. It is not a *uniform exaltation* of the mental powers, but in some degree a *perversion* of them; for that voluntary control over the current of thought, which is the distinguishing character of the sane mind of Man, is considerably weakened, so that the heightened imagination and enlivened fancy have more unrestricted exercise; and whilst ideas and images succeed each other in the mind with marvellous readiness, no single train of thought can be carried out with the same continuity as in the state of perfect sobriety. This weakening of the voluntary control over the mental operations must be regarded, then, as an incipient stage of Insanity.

13. If the first dose of Alcohol be such as to produce more potent effects, or if (as in ordinary intoxication) it be renewed after the first effects have been already manifested, the *second* stage is induced, in which not merely the intellectual but the sensorial apparatus[3] is disturbed. The voluntary control over the direction of the thoughts is completely lost, and the excitement has more the character of

---

[1] *In vino veritas*—there is truth in wine. The influence of wine is to develope the true character of the individual.

[2] *Phrenitis*—inflammation of the brain.

[3] *Sensorial apparatus*—that portion of the nervous system upon which our sensations, feelings, perceptions, depend.

delirium ;—the ideas becoming confused, the reasoning powers dis. ordered, and hallucinations[1] sometimes presenting themselves. At the same time, vertigo, double vision, tinnitus aurium,[2] and various other sensory illusions occur; the muscular movements become tremulous and unsteady, the voice thick, the eyes vacant, and the face commonly pale. Vomiting frequently occurs in this state; and when it does, the consecutive stage is usually either cut short, or is abated in intensity. The poisonous effects may proceed no further than this; the drunkard falling into a heavy sleep, from which he awakes to feel the consequences of his transgression. These consequences differ in some degree with the previous habits. Those unaccustomed to such excesses usually suffer from head-ache and feverishness, with a dry and furred tongue, complete anorexia[3] with a particular loathing for alcoholic drinks, inability for mental or bodily exertion, and depression of spirits; and only recover from this condition after prolonged repose and abstinence. On the other hand, the man to whom it is habitual, although his general condition is nearly the same, craves for a further dose of his stimulant; and when he has obtained it, he is able to take food and to proceed with his ordinary avocations.

14. In the *third* and most profound stage of intoxication, there is extreme diminution or entire suspension of cerebral and sensorial power;[1] a state of *coma*[2] supervening upon that last described. This state may vary in intensity, however, between one of deep ordinary sleep, from which the individual can be so far aroused as to give manifestations of sensibility, and a torpor as profound as that of apoplexy; and, when the latter manifests itself, it is an indication of danger, especially when the respiratory movements are impeded. According to the observations of Dr. Ogston,* the face is sometimes pale, sometimes flushed; the eyes vacant and suffused, sometimes glazed; the pupils dilated, and contracting very imperfectly, or not at all, to light: the temperature of the head is generally above the natural standard, but that of the extremities and of the surface is in general considerably lowered, or but little affected in milder cases : the pulse, which was at first quick and excited, becomes feeble, small, and ultimately slow, or even entirely wanting at the wrist, according to the intensity of the intoxication; the respiratory movements are less frequent than usual, and are imperfectly performed, exhibiting, in the severest cases, the convulsive character of those of persons

---

[1] *Hallucinations*—errors of the senses, in which objects or impressions are supposed to be seen, felt, or heard, that are not in fact present.

[2] *Tinnitus aurium*—a ringing in the ears.

[3] *Anorexia*—a want or loss of appetite; aversion from food.

[4] *Cerebral and sensorial*—relating to the brain or mind and senses.

[5] *Coma*—a state of deep sleep.

Edinburgh Medical and Surgical Journal, vol. xl.

suffering from asphyxia.[1] Strabismus,[2] general tetanic convulsions,[3] or spasms of particular parts, sometimes supervene in the more advanced states. When a fatal termination occurs, it is usually attributable, as in apoplexy, to the imperfect aeration of the blood; the face becoming livid and tumid, the eyes prominent, and the lips blue. In some instances, the complete prostration of the cerebral and sensorial powers comes on suddenly, without any previous stage of excitement; and in these cases, it is noticed that the pupil is usually contracted.

15. The unfavourable indications, in case of poisoning by large doses of Alcohol, are profoundness of insensibility, insufficiency of respiratory movement, with consequent failure of circulation and imperfect aeration[4] of blood, the pupils either much dilated or contracted, coldness of the extremities, and the occurrence of strabismus or tetanic spasms. When these symptoms do not appear, the ill effects pass off, in a great measure, within four-and-twenty hours; but it is usually much longer before the various functions regain their healthy tone.

16. In fatal cases,[5] the appearances usually resemble, more or less closely, those of Asphyxia; the right side of the heart, the pulmonary arteries, and the systemic veins[6] being loaded with blood; whilst the left cavities and the arterial system are comparatively empty, the blood which they do contain being dark. The sinuses and the whole venous system of the brain are turgid with dark blood; and serous effusion[7] is usually found within the ventricles,[8] or beneath the arachnoid;[9] this, however, being variable in its amount. The substance of the brain is unusually white and firm, as if it had lain in alcohol for an hour or two. The liver, spleen, and kidneys are loaded with venous blood; and the air-passages of the lungs contain more or less of frothy mucus. The stomach usually exhibits but little departure from its normal condition, except in cases where drunkenness has been habitual, or where the fatal dose has been taken in a very concentrated form. In the former case, the mucous

---

[1] *Asphyxia* — interruption of respiration or the breathing function — strangulation, suffocation.

[2] *Strabismus* — squinting.

[3] *Tetanic convulsions* — convulsions in which the muscles are permanently extended or contracted; continued spasms.

[4] *Aeration* — change of the blood by the action of the air in the lungs

[5] See Dr. Ogston, *loc. cit.*; and Dr. Peters in New York, Journal of Medicine, vol. iii. no. 9.

[6] *Pulmonary arteries and systemic veins* — vessels which return the blood, after it has circulated through the body, to the lungs, to be again aerated.

[7] *Serous effusion* — a pouring out or escape of the watery portion of the blood.

[8] *Ventricles* — cavities within the brain.

[9] *Arachnoid* — one of the membranes enveloping the brain.

coat[1] is usually found thicker, softer, and more vascular than usual; this change sometimes extending even throughout the entire length of the small intestines. In rarer instances, the coats of the stomach are remarkably thickened and hardened. Where death results from a very concentrated dose, the intense injection, almost amounting to black discolouration, of a powerful irritant poison, is met with. This, however, has been rather noticed in experiments on animals, into whose stomachs rectified alcohol had been injected, than in human subjects, by whom alcohol is very rarely taken in such a form. Of the condition of the liver and kidneys found in *habitual* drunkards, an account will be given hereafter. The blood in most cases of alcoholic poisoning, according to the observations of Dr. Ogston upon drunkards, and the experiments of Dr. Percy upon animals, is either fluid or imperfectly coagulated.

Pathology of Alcoholic Intoxication.

17. The *pathological* character of intoxication by Alcohol, and the *modus operandi* of the poison, have been fully made out from the experiments and observations just referred to; and it is very important for our future inquiries, that the results of these should be rightly understood. That alcoholic liquors, when introduced into the stomach, should undergo rapid absorption into the sanguiferous system,[2] is precisely what might be anticipated from our knowledge of the conditions under which that absorption takes place; and there is ample evidence that such is really the case. Thus Dr. Percy was always able to detect the alcohol in the blood of the animals which he had poisoned by injecting alcohol into their stomachs, provided they did not live too long afterwards; and MM. Bouchardat and Sandras have more recently determined its presence in the blood of the gastric veins.[3] The rapidity with which this absorption takes place may be judged of by the fact that, in one of Dr. Percy's experiments, in which the animal fell lifeless to the ground immediately that the injection of the alcohol into the stomach was completed (the respiratory movements and pulsations of the heart entirely ceasing within two minutes), the stomach was found nearly void, whilst the blood was strongly impregnated with alcohol.[4] Hence, it may reasonably be inferred, that in all cases of ordinary intoxication, and in the greater number of cases of death from the introduction of alcohol into the stomach, the effects are produced by the passage of the alcohol into the current of the circulation, so as to exert a direct action on the nervous centres. And this influence is confirmed by the fact that Dr. Percy has demonstrated its pre-

---

[1] *Mucous coat* — inner or lining coat of the stomach.
[2] *Sanguiferous system* — The blood-vessels, veins and arteries.
[3] *Gastric veins* — veins of the stomach.
[4] Op. cit. p. 61.

sence, in considerable amount, *in the substance of the brain;* thus confirming Dr. Ogston's assertion (which had been called in question by Dr. Christison and others) as to its presence in the fluid effused in the ventricles of the brain.

18. In some of the experiments on Alcoholic poisoning, however, made by Dr. Christison and others, it would appear as if the total loss of insensibility and voluntary power so *instantaneously* followed the introduction of the poison into the stomach—especially when it was introduced in a concentrated form—as not to admit the idea that absorption could have taken place to a sufficient extent for the production of the effect by the direct action of the poison on the nervous centres. In such instances, the fatal result would seem rather due to the violent impression made upon the gastric nerves, especially those of the sympathetic system; whereby the heart's action is suspended, and death takes place by *syncope*[1] rather than *asphyxia*.[2] This is the case with many other poisons, when administered in large quantity and in concentrated form, especially with such as exert a chemical action upon the animal tissues; the effect which they produce (through the nervous system) upon the heart, closely resembling that of blows upon the epigastrium,[3] or extensive burns of the cutaneous surface.[4] Now as the concentration of the alcohol will, on the one hand, favour its physical and chemical action upon the tissues; whilst on the other it will be unfavourable to absorption, which takes place much more readily when it is diluted with water; we are enabled readily to account for this difference in its *modus operandi.*

19. The *general* stimulant action, which is for a time exerted by alcohol introduced in small quantities, and diluted by admixture with the general mass of the blood, is easily explained upon the basis of the observations first detailed (§ 17); but its special power of exciting the *nervous centres*[5] to augmented activity, can only be accounted for by the idea of some special relation between alcohol and nervous matter. And this idea is fully borne out by the fact, that Dr. Percy found alcohol to exist in the substance of the brains of the dogs poisoned by it, in considerably greater proportion than in an equivalent quantity of blood. (*Op. cit.* p. 103.)—This fact is one of fundamental importance, as showing us how directly and immediately the whole nutrition and vital activity of the Nervous System must be affected by the presence of Alcohol in the blood; the alcohol being thus specially drawn out of the circulating current

---

[1] *Syncope* — fainting; cessation of the heart's action.
[2] *Asphyxia* — stoppage of respiration; cessation of the function of the lungs.
[3] *Epigastrium* — region of the stomach.
[4] *Cutaneous surface* — the skin.
[5] *Nervous centres* — the brain and spinal marrow.

by the nervous matter, and incorporated with its substance, in such a manner as even to change (when in sufficient amount) its physical as well as its chemical properties. It is important also to observe, that this affinity is obviously such, as will occasion the continual presence of alcohol in the blood, even in very minute proportion, to modify the nutrition of the *nervous* substance more than that of any other tissue; for the alcohol will *seek out* (as it were) the nervous matter, and will fasten itself upon it, — just as we see that other poisons, whose *results* become more obvious to our senses, (although the poisons themselves may exist in such minute amount as not to be detectible by the most refined analysis,) will localize themselves in particular organs, or even in particular spots of the same organ.[1]

20. The selective power of Alcohol appears to lead it in the first instance to attack the *Cerebrum*,[2] the intellectual powers being affected before any disorder of sensation or motion manifests itself; and to this it seems to be limited in what has been here described as the *first* stage of intoxication. But with the more complete perversion of the intellectual powers, which characterises the *second* stage, we have also a disturbed function of the *Sensory Ganglia*,[3] upon which the cerebral hemispheres[4] are superposed; this disturbance being indicated by the disorders of sensation, and also by the want of that control over the muscular movements which require sensation for their guidance. In the *third* stage, the functions of the Cerebrum and Sensory Ganglia appear to be completely suspended; and those of the *Medulla Oblongata*[5] and *Spinal Cord*[6] now begin to be affected, as we see to be indicated by the difficulty of respiration, the strabismus, the dilated pupil, and the tetanic spasms. As already stated, the admixture of alcohol with the blood has a tendency to give a venous character even to that of the arteries; and when this tendency is augmented by imperfect respiration, the blood will become more and more venous, until its influence upon the medulla oblongata is so directly poisonous, that its functions are completely suspended, the respiratory movements are brought to a stand, and death takes place by Asphyxia, precisely as in narcotic poisoning by other substances.

21. This tendency, however, is more completely antagonized by

---

[1] See for illustrations of this doctrine, now generally admitted by physiologists, Dr. W. Budd's paper on Symmetrical Diseases, in *Med. Chir. Trans.*, vol. xxv.; and Mr. Paget's Lectures on Nutrition, in *Medical Gazette*, 1847.

[2] *Cerebrum* — the upper and anterior portion of the brain.

[3] *Sensory ganglia* — portions of the nervous system from which the nerves of sensation proceed.

[4] *Cerebral hemispheres* — the two lateral portions of the brain.

[5] *Medulla oblongata* — the commencement of the spinal marrow at the basis of the brain within the scull.

[6] *Spinal cord* — the spinal marrow.

the efforts which the system makes (so to speak) to get rid of the poison; which efforts, if successful, will carry it off in the course of a few hours, leaving nothing behind it but the disordered condition which it has produced. We have seen that an increased secretion [1] takes place by the kidneys and skin; and the former of these is certainly a means of eliminating [2] the alcohol, which has been detected in the *urine* by Dr. Percy, (*op. cit.*, p. 104,)—contrary to the statements of many physiologists, who have denied that it ever finds its way into that secretion. It is indeed a general rule, that when a medicinal or toxic [3] agent produces a special determination to some particular gland, that determination is the means of eliminating it from the blood; as is seen in the diuretic action of the neutral salts. And it would not seem improbable, therefore, that the *skin* also should be concerned in the exhalation [4] of the alcohol; [5] more especially since an alcoholic odour may often be observed, not merely in the breath, but proceeding from the person generally. Dr. Percy has also shown, that alcohol may be detected in the *bile* of animals poisoned by it. The alcoholic odour of the *breath* is a sufficient indication that alcoholic vapour is exhaled from the lungs in the act of respiration;—but the quantity of this is probably small in comparison with that which is carried off in another way, namely, by the combustive process, which takes place in the blood at the expense of the oxygen it contains, and which converts the alcohol into carbonic acid and water; both of which are set free by exhalation from the lungs. The readiness with which alcohol is thus oxidized, in fact, is probably one cause of its influence in giving a venous aspect to arterial blood; since it will withdraw the oxygen from other substances, which are waiting to be eliminated by the combustive process, and the accumulation of which will deteriorate the character of the fluid.

22. By all these channels, then, the Alcohol is rapidly carried out of the system; so that recovery may be certainly expected, if life can be sufficiently prolonged by warmth to the surface, by artificial respiration, by the administration of ammonia, [6] and by other such measures. The *depressing* character of the influence of alcohol, when carried to this extent, is sufficiently indicated by the fact

---

[1] *Increased secretion by the kidneys and skin* — increases separation from the blood of urine and sweat.

[2] *Eliminating* — separating, removing.

[3] *Toxic* — poisonous.

[4] *Exhalation* — giving out in the form of a fine vapour.

[5] Dr. Macnish states, (Anatomy of Drunkenness, p. 175), that he has met with two instances, the one in a Claret, the other in a Port drinker, in which the cutaneous perspiration after a debauch had the hue of the liquor they had drunk.

[6] *Ammonia* — volatile alkali, hartshorne.

4

that copious depletion [1] cannot be borne; and it is on this account very important to distinguish between alcoholic poisoning and congestive apoplexy, for which it may be easily mistaken if its cause be not known.

It would be easy to extend the foregoing description by a more particular account of the varieties of the *modus operandi*[2] of Alcohol presented by different individuals; but it has not been thought necessary to do so, the great general facts presented by its ordinary operation being those of prime importance in our further investigations.—We have now to inquire into the various forms of *disease*, the production of which may be assigned, with more or less probability, to the prolonged or repeated action of alcohol on the human system.

III. REMOTE CONSEQUENCES OF THE EXCESSIVE USE OF ALCOHOLIC LIQUORS.

Diseases of the Nervous System.

23. From the peculiar tendency which the Alcohol in the blood has to disorder the functions of the Nervous System, it might be expected that the habitual ingestion of alcoholic liquors in excess, would bring about a more permanent derangement of this apparatus; and more particularly of its Encephalic[3] portion,—which seems to be singled out by alcohol, almost to the exclusion of the spinal cord, just as the spinal cord is affected by strychnine,[4] almost to the exclusion of the encephalon.[5] This we shall find to be the case. There are, in fact, scarcely any diseases of the Encephalon, except such as are of a purely constitutional nature, (such as tubercular[6] or cancerous affections,) which are not so much more frequent among the habitually intemperate than among the habitually sober, as to justify us in regarding the excessive use of alcoholic liquors as among the most efficacious of the conditions of their production. It will be proper, therefore, to pass the principal diseases in review before us, and to inquire into the mode in which habitual excess in the use of alcoholic drinks tends to produce each of them.

24. *Delirium Ebriosum.*[7]—Considering that the state of intoxication is itself, strictly speaking, a transient paroxysm of insanity, it can excite no surprise that a confirmed state of mental derange-

---

[1] *Depletion* — removal or loss of blood or of its constituents by bleeding, purging, &c.

[2] *Modus operandi* — mode or manner in which it acts.

[3] *Encephalic portion* — the portion within the head.

[4] *Strychnine* — the active principle of nux vomica.

[5] *Encephalon* — the brain.

[6] *Tubercular* — scrofulous, consumptive.

Delirium ebriosum — the delirium of drunkenness.

ment should frequently result from the repetition of the cause which produces the single paroxysm. There are, in fact, some individuals in whom a fit of positive madness, persisting for some little time after the immediate effects of the stimulus have subsided, is brought on by every excess in drinking. The head becomes extremely hot, the face flushed, the pulse very frequent, full, and hard, the temper is excessively violent, the individual sometimes attacking every one who comes in his way, and being always prone to ferocity against any one who opposes him; and all sense of danger being lost, he is not deterred from violence by the fear of personal injury, but rushes madly upon what may prove his destruction. This condition, the *delirium ebriosum* of Darwin, is obviously an exaggeration of one of the ordinary forms of excitement in common intoxication; and it usually subsides in a day or two, if the individual be simply restrained from doing mischief to himself or others. It is sometimes accompanied, however, with *tremors*,[1] even in the midst of violent excitement; and this form constitutes the transition to the disorder next to be noticed. The frequent repetition of this paroxysm, of which, as of ordinary drunkenness, the stimulating action of alcohol on the nervous centres must be regarded as the immediate cause, is almost certain, like the recurrence of regular maniacal paroxysms, to end in some settled form of Insanity.

25. *Delirium Tremens*.[2]—The habitual drunkard, who has exhausted his nervous power by continual over-excitement, is liable to another form of disordered action of his brain, which is commonly known from one of its most marked symptoms,—the peculiar tremor of the limbs,—as *delirium tremens*. This state is in many respects the opposite of the preceding. There is little or no heat of the head or flushing of the face, the skin is cool and humid, and even chilly; the pulse, though frequent, is small and weak; and the temper, though very irritable, is not violent,—the prominent disposition, indeed, being *anxiety* and *apprehension* of injury or danger. There is an almost entire want of sleep; and even if repose be obtained, it is very imperfect, being interrupted by frightful dreams. On the other hand, the waking state is frequently so disturbed by illusions of a disagreeable or frightful nature, that it differs but little from that of sleep, save in the partial consciousness of external things. The following is the vivid picture of this condition, given by one who has himself experienced it.[3] "For three days I endured more agony than pen could describe, even were it guided by the hand of a Dante. Who can tell the horrors of that horrible malady, aggravated as it is by the almost ever-abiding consciousness that it is self-sought? Hideous faces appeared on the walls, and on the ceil-

---

[1] *Tremors* — an unsteady, tremulous motion of the limbs; shaking.
[2] *Delirium tremens* — delirium with tremors of the limbs; mania à potu.
[3] Autobiography of J. B. Gough, p. 70.

ing, and on the floors; foul things crept along the bed-clothes, and glaring eyes peered into mine. I was at one time surrounded by millions of monstrous spiders, who crawled slowly, slowly over every limb; whilst beaded drops of perspiration would start to my brow, and my limbs would shiver until the' bed rattled again. Strange lights would dance before my eyes, and then suddenly the very blackness of darkness would appal me by its dense gloom. All at once, whilst gazing at a frightful creation of my distempered mind, I seemed struck with sudden blindness. I knew a candle was burning in the room, but I could not see it. All was so pitchy dark. I lost the sense of feeling too, for I endeavoured to grasp my arm in one hand, but consciousness was gone. I put my hand to my side, my head, but felt nothing, and still I knew my limbs and frame *were* there. And then the scene would change. I was falling— falling swiftly as an arrow far down into some terrible abyss; and so like reality was it, that as I fell I could see the rocky sides of the horrible shaft, where mocking, gibing, mowing, fiend-like forms were perched; and I could feel the air rushing past me, making my hair stream out by the force of the unwholesome blast. Then the paroxysm sometimes ceased for a few moments, and I would sink back on my pallet drenched with perspiration, utterly exhausted, and feeling a dreadful certainty of the renewal of my torments."

26. With this disturbed condition of the brain, a more or less disordered state of the digestive apparatus is commonly associated. The tongue is furred, the stomach unable to bear food without vomiting or a sense of oppression; the bowels are usually constipated, or, if they be relaxed, the stools are dark and offensive, and the urine is scanty. Sometimes the disease proceeds to a fatal termination, which is generally preceded by contracted pupil, occasionally strabismus, incessant low delirium, increase in the muscular tremor almost amounting to subsultus tendinum,[1] and other indications of nervous exhaustion; the pulse becomes thready, and at the same time more rapid, so that it sometimes can scarcely be counted; cold sweats break out upon the skin, and the chillness of the surface increases, proceeding from the extremities to the trunk. Sometimes a calm supervenes shortly before death; whilst in other instances the patient is carried off in a convulsion. On the other hand the bad symptoms may gradually abate, and the supervention of profound sleep gives to the exhausted energies of the nervous system the means of restoration. Sometimes, however, the recovery is never complete, but the patient remains in a state of Melancholia, with more or less of deficiency of intellectual power; and this more especially happens after repeated attacks of the disease.

27. Between the *Delirium Ebriosum* and the proper *Delirium*

---

[1] *Subsultus tendinum* — a starting or spasmodic twitching of the tendons.

Tremens, there are several intermediate conditions; the former, indeed, being very apt to pass into the latter, if depleting measures[1] be imprudently adopted. The latter may present itself, like the former, as the direct and immediate consequence of the excessive use of Alcoholic liquors; but *there is this important difference,— that while the former is but an exalted manifestation of the primary *excitement* ordinarily produced by alcohol, the latter, supervening at the end of a prolonged debauch, is the consequence of *exhaustion* produced by continued excitement. Delirium Tremens more frequently occurs, however, when the accustomed stimulus is withheld; and it is then no less obviously the result of the previously *exhausted* condition of the nervous system, which nothing save the renewal of the potent stimulus can excite to anything like regular action. In fact, this terrible state is the manifestation of the disordered condition to which the brain has been brought by habitual excess, and plainly exhibits the complete perversion of its functional power and of its nutritive operations. In fatal cases, no morbid appearances are found, that in the least indicate turgescence of the vessels or inflammatory excitement, unless the delirium have partaken of the characters of that which directly arises out of intoxication. And it is manifest, therefore, that the disordered condition must be in the nervous pulp itself, and that it must be of a kind to keep up morbid and irritative activity, at the same time that the tissue is incapable of exercising those reparative functions, which are carried on in the healthy condition during the state of repose.

28. Although, in the vast majority of cases, Delirium Tremens is the immediate or the consecutive result of the excessive use of Alcoholic liquors, yet it may occur independently of them; but its other causes are such as resemble the excitement of alcohol, in producing *exhaustion* or *depression* of the nervous power,—such, for instance, as excessive depletion, the shock of severe injuries, or extreme cold. But in most of the cases in which one or other of these appears to be the exciting cause, a predisposition has been established by habitual intemperance; and this has been especially remarked of the *delirium traumaticum*.[2]

29. It is important to remark, that a slighter form of this disorder, marked by tremors of the hands and feet, deficiency of nervous power, and occasional illusions, will sometimes appear as a consequence of habitual tippling, even without intoxication having been once produced. And a still slighter manifestation of the want of control over the muscular apparatus—the trembling of the hands in the execution of a voluntary movement—is familiar to every one as extremely frequent among the habitually intemperate. We thus

---

[1] *Depleting measures* — whatever empties the blood-vessels; bleeding, purging, &c.

[2] *Delirium traumaticum* — delirium from wounds or injuries.

4 *

see that the disease is at least as much dependent upon the *disor-dered state of nutrition*, consequent upon the habitual presence of alcohol in the blood, as it is upon that positive exhaustion of nerv-ous power consequent upon the violence of the excitement, which is the more immediate effect of the stimulus.

30. *Insanity.* — Such being the case, we have no difficulty in understanding how the habitual use of Alcoholic liquors in excess becomes one of the most frequent causes of *Insanity*, properly so called, *i. e.* of settled Mental Derangement.   Upon that point, all writers on the subject are agreed, however much they may differ in their appreciation of the relative frequency of this and of other causes.   The proportion, in fact, will vary according to the character of the population on which the estimate has been formed; and also according to the mode in which it has been made.   Thus, in Pauper Lunatic Asylums, the proportion of those who have become insane from Intemperance, is usually much larger than it is in Asylums for the reception of Lunatics from the higher classes, among whom intemperance is less frequent, while causes of a purely moral and intellectual nature operate upon them with greater inten-sity.   And again, if in all cases in which habitual intemperance has been practised, it be set down as the cause of the mental disorder, the proportion becomes much larger than it will be, if (as happens in many cases,) some other cause have been in operation concur-rently, and the disorder be set down as *its* result, no notice what-ever being taken of the habit of intemperance.   This omission must be particularly allowed for, when the relative proportion of intem-perance to other causes is being estimated in regard to the middle and higher classes; on account of the strong desire which usually exists among the friends of the patient, to conceal the nature of his previous habits, and to lay his disorder entirely to the account of the cause from which it has seemed immediately to proceed.

31. There can be no doubt that those who have weakened and disordered the nutrition of the brain by habitual Intemperance, are far more liable than others to be strongly affected by those causes, moral or physical, to which the Mental Derangement is more imme-diately attributable; so that the habit of intemperance has con-tributed, as a *predisposing* cause, at least as much towards its pro-duction, as what is commonly termed the *exciting* cause has done. In fact, of predisposing causes generally, it may be remarked, that their action upon the system is that of slowly and imperceptibly modifying its *nutritive* operations, so as gradually to alter the che-mical, physical, and thereby the vital properties of the fabric; and thus to prepare it for being acted on by causes which, in the healthy condition, produce no influence.   And although *that one* of the conditions in previous operation is often singled out as *the cause*, from which the result may seem most directly to proceed, yet it

frequently happens that it has really had a far smaller share in the production of the disorder, than those remoter causes whose operation has been more enduring and really more effectual.[1]

32. In the Statistical Tables, published by the Metropolitan Commissioners of Lunacy, in 1844, comprehending the returns from 98 Asylums in England and Wales, we find that out of 12,007 cases whose supposed causes were returned, 1799, or nearly 15 per cent., are set down to the account of Intemperance; but besides these, 551 or 4·6 per cent. are attributed to vice and sensuality, in which excessive use of alcoholic liquors must have shared. Moreover in every case in which Hereditary Predisposition was traceable, *this* was set down as *the* cause; notwithstanding the notorious fact that such predisposition frequently remains dormant until it is called forth by habitual intemperance. It is not more correct, therefore, to regard this as *the* cause of the disorder, in all the cases in which it is traceable, than it would be to regard intemperance in that light, in every case in which the patient had previously indulged in alcoholic excesses. Of the 2526 cases, then, in which the disorder is attributed to hereditary predisposition, a considerable proportion might with equal justice be set down to the account of intemperance. And there can be no doubt that the same practice had a great share in the production of the disease in the 3187 cases set down to bodily disorder, and in the 2969 for which moral causes are assigned.

33. If we turn from this general statement to the experience of individual Asylums, we frequently find the proportion much higher; and curious variations are sometimes observable between the returns for successive years. Thus in the Glasgow Lunatic Asylum, according to the report of Dr. Hutcheson, the following were the proportions which Intemperance bore to other causes during seven years :—

Year.	Total number of patients.	Cases where the disease was hereditary, &c.	Cases where the cause was unknown.	Cases where intemperance was the cause.	Proportion per cent. of intemperance to other causes.
1840	149	3	34	20	13·4
1841	157	20	44	30	19·1
1842	199	54	20	46	23·1
1843	327	116	38	31	9·4[2]
1844	390	77	41	53	18·2
1845	364	47	38	90	24·7
1846	414	49	62	105	25·3
Total	1900	366	277	375	19·7

[1] See Mr. J. S. Mill's Elements of Logic, vol. i. p. 398.
[2] This marked diminution in the per centage of cases attributable to Intemperance is chiefly due to the admission into the Glasgow Asylum

Of the great increase which presents itself in the number of cases attributable to intemperance during the last two years of this return, Dr. Hutcheson thus speaks in his report for 1846 :—"This cause appears to have operated on patients of all ranks; and I am inclined to think that this has been owing, in a great measure, to the excitement in which the community was kept, by that universal spirit of gambling which seized on society like an epidemic mania. There is a great connection between general excitement and the craving for stimulants, as may be every day seen during contested elections, public dinners, races, &c.   It is also a fact well known to those who have minutely studied the subject, that over-exertion of the brain leads to a desire for stimulants, which, however, are easily enough abandoned when the brain is allowed to rest." For the reason already given, it is probable that the average proportion of 19·7 per cent. does not by any means represent the entire number of cases in which intemperance was the principal cause of the disease; and that we should be within the truth in assigning to it at least *a quarter* of the whole number of cases.

34. In the report of the Aberdeen Lunatic Asylum for 1847, we find Intemperance specified as the cause in 17 cases out of 93 admitted; but of these 93, there were 8 cases in which hereditary predisposition, and 11 in which predisposition from previous attacks, was assigned as the cause; and there can be no doubt that of these 19 cases a considerable proportion might be set down in part to the account of intemperance.   In the report of the Dundee Lunatic Asylum, we find that 8 out of 52 cases admitted are set down to intemperance; 7 were hereditary; and in 4 the cause was unknown.   In other Asylums, the proportion of cases returned as

---

during the year 1843, of a number of Lunatics who had previously been confined at Arran, for the most part during several years.   Of the origin of their Insanity very little was known; and they were chiefly assigned to the head of "Hereditary and Constitutional Predisposition," thereby diminishing the per centage of the other causes.   Among the recent cases admitted during the year, however, the per centage attributable to intemperance was decidedly less than usual; which circumstance is attributed by Dr. Hutcheson to the improved condition of trade, which caused an adequate demand for labour.   On this point he makes the following remarks, in his report for 1842, p. 36 :—"It may be said, that when wages ... e low and occupation difficult to be obtained, men will have less money to spend, and consequently will drink less.   A pretty extensive observation of the different grades of the working classes, for upwards of fifteen years, has convinced me that this opinion is erroneous; for I have generally found, that want and intemperance go hand in hand.   Whenever a man falls below a certain point in physical comfort, he becomes reckless, and sensual enjoyment forms his only pleasure.   To this he will sacrifice everything; and habits of intemperance are frequently acquired in seasons of distress, which the individual, in more favourable circumstances, finds it impossible to lay aside."

due to Intemperance is much greater than in those already referred to. Thus in the Commissioner's report already cited, we find that in nine Provincial private Asylums, the proportion which the cases assigned to Intemperance alone bears to those assigned to other causes, is no less than 32·62 per cent.; independently of 5·67 per cent. which are set down to the account of "Vice and Sensuality." There is an Asylum in the East of London, where the proportion of cases attributed to intemperance alone amounted to 41·07 per cent.; and those arising out of this in combination with other vices, to 22 per cent. of the whole number whose causes were assigned. And it is stated by Dr. Macnish, (*op. cit.*, p. 193,) that of 286 Lunatics at that time in the Richmond Hospital, Dublin, *one half* owed their madness to drinking.

35. *Oinomania.*[1] — There is one form of Insanity, which has so peculiar a relation to the use of alcoholic liquors, as to call for particular consideration in this place; and in order that its characters may be presented in the most unexceptionable manner, the author avails himself of the excellent account of the disease which is given by Dr. Hutcheson in the Report of the Glasgow Lunatic Asylum for 1842, (pp. 39–44); deeming its value sufficient to justify him in presenting it without abridgment. The designation *Oinomania*, he would remark, has been substituted by Dr. H. for the less appropriate term *Dipsomania* used by other authors. "This form of mania," he observes, "is quite different from drunkenness, which, however, may lead to it; the diagnostic[2] mark of the disease being the irresistible propensity to swallow stimulants in enormous doses, whenever and wherever they can be procured. There are individuals who at the festive board invariably become excited, if not intoxicated, but who are otherwise habitually sober, and in the course of the year drink much less than others who never appear to be under the influence of stimulants. Others indulge in their potations in a regular manner, and daily consume a larger quantity of liquor than is consistent with good health or sobriety. All these, however, possess self-control, and can at any time refrain from stimulants; but those affected with the disease cannot do so, however convinced they may be of the impro7priety of yielding to their propensity, or however desirous they may be to subdue it. I repeat, that the disease does not consist in the mere act or habit of becoming intoxicated; but in the irresistible impulse which drives the unhappy being to do that which he knows to be pernicious and wrong, and which, in the intervals of his paroxysms, he views with loathing and disgust. He derives no pleasure from taste, for he gulps down the liquor of whatever kind it may be; or from society,

---

[1] *Oinomania* — wine mania; an inordinate and uncontrollable thirst for excessive quantities of intoxicating drinks.

[2] *Diagnostic* — distinguishing.

for he generally avoids society; but he only derives a temporary satisfaction from the gratification of his insane impulse, or rather from freeing himself from the overwhelming misery which the non-gratification of his impulse inflicts on him. The disease appears in three forms—the acute, the periodic, and the chronic.

"The *Acute* is the rarest of the three. I have seen it occur from hemorrhage in the puerperal state, in recovery from fevers, from excessive venereal indulgence, and in some forms of dyspepsia. When it proceeds from any of the first three causes, it is easily cured by restoring the strength of the patient. When it arises from the fourth cause mentioned, it is not so easily removed, and is very apt to assume the chronic form.

"The *Periodic* or paroxysmal form is much more frequent than the acute. This is often observed in individuals who have suffered from injuries of the head, females during pregnancy, at the cata-menial periods, on the approach of the critical period and after-wards, and in men whose brains are overworked. When it occurs from injury of the head, the case is hopeless. In the other in-stances it may be cured. In some cases it occurs whenever the individual partakes of stimulants. In these, total abstinence is the only remedy. Like the form about to be mentioned, it is frequently hereditary, being derived from a parent predisposed to insanity or addicted to intemperance. In such cases, the probability of cure is very small. The individual thus affected abstains for weeks or months from all stimulants, and frequently loathes them for the same period. But by degrees he becomes uneasy, listless, and de-pressed, feels incapable of application, and restless, and at last begins to drink till he is intoxicated. He awakes from a restless sleep, seeks again a repetition of the intoxicating dose, and con-tinues the same course for a week or longer. Then a stage of apathy and depression follows, during which he feels a loathing for stimulants, is the prey of remorse, and regrets bitterly his yielding to his malady. This is followed by fresh vigour, diligent applica-tion to business, and a determined resolution never again to give way. But, alas! sooner or later the paroxysm recurs, and the same scene is re-enacted, till ultimately, unless the disease be checked, he falls a victim to the physical effects of intemperance, becomes maniacal, or imbecile, or affected with the form of the disease next to be mentioned.

"Of all the forms of *Oinomania* the most common is the *chronic*. The causes of this are injuries of the head, diseases of the heart, hereditary predisposition, and intemperance. This is by far the most incurable form of the malady. The patient is incessantly under the most overwhelming desire for stimulants. He will dis-regard every impediment, sacrifice comfort and reputation, with-stand the claims of affection, consign his family to misery and disgrace, and deny himself the common necessaries of life to gratify his insane propensity. In the morning, morose and fretful, disgusted with himself, and dissatisfied with all around him, weak and tremulous, incapable of any exertion either of mind or body, his first feeling is a desire for stimulants, with every fresh dose of which he recovers a certain degree of vigour both of body and mind, till he feels comparatively comfortable. A few hours pass without the craving being so strong; but it soon returns, and the patient drinks till intoxication is produced. Then succeed the restless sleep, the suffering, the comparative tranquillity, the excitement, and the state of insensibility; and, unless absolutely secluded from all means of gratifying the propensity, the patient continues the same course till he dies, or becomes imbecile. This is that fearful state portrayed by Charles Lamb, in which reason revisits the mind only during the transient period of incipient in toxication.

" It must be remarked, that in all these forms of the disease the patient is perfectly incapable of self-control; that he is impelled by an irresistible impulse to gratify his propensity; that while the paroxysm is on him, he is regardless of his health, his life, and all that can make life dear to him; that he is prone to dissipate his property, and easily becomes the prey of the designing; and that in many cases he exhibits a propensity to commit homicide or suicide. He is thus dangerous to himself and others, and however responsible he may have been for bringing the disease on himself, his responsibility ceases as soon as he comes under the influence of the malady. The disease, however, may not be brought on by the act of the individual; and then it is clear at once, that neither directly nor indirectly can he be deemed responsible. But suppose that it were the result of his previous conduct, I repeat that however culpable he may have been for that, he is not a responsible being while afflicted with the malady; for I can see no distinction between this form of the disease and any other which has been in duced by the habits or acts of the individual.

" The only chance of cure or alleviation, is from attention to the health, and abstinence from intoxicating liquors. Neither can be secured so long as the patient is at large; and no amendment can be depended on, unless he has undergone a long course of discipline and probation. Considering, then, that the individual is irresponsible and dangerous to himself and others—that, if left uncontrolled, he will ruin his family—and that his disease can be treated only in an Asylum, it is not only merciful to him and his relatives, but necessary for the security of the public, that he be deprived of the liberty which he abuses and perverts, and that he should be prevented from committing crimes instead of being punished, or I should rather say, being the object of vindictive infliction after he has per-

petrated them.   So convinced are some affected with the periodical
form of the disease, of the necessity of being controlled, that, when
the first symptoms of their paroxysm are felt, they voluntarily enter
an Asylum, and remain till the attack has passed off.   These, however, are men of stronger minds, though, with all their strength,
incapable of resisting the disease; and, surely, what they feel to be
their only refuge to avoid the impending evil, it cannot be unjust or
harsh to force on others whose minds are more impaired.   Such
cases soon become rational in an Asylum; and when the individual
can so far control himself as voluntarily to surrender his liberty on
the first premonitory symptoms of the malady presenting themselves,
he may be dismissed after a shorter probation.   It is otherwise with
those who have not that self-control, or who fancy that they are un-
justly interfered with when checked in their career.   They require
a much longer probation, which should be increased at each return
of their malady.

"Of the chronic form, I have seen only one case completely cured,
and that after a seclusion of two years' duration.   In general it is
not cured; and no sooner is the patient liberated, than he manifests all the symptoms of his disease.   Paradoxical though the statement may appear to be, such individuals are sane only when confined
in an Asylum."

The Superintendent of the Dundee Asylum, in remarking upon
the frequent causation of Insanity by Intemperance, makes a very
similar statement of the results of his observations; and regrets that
there are not in this country such Asylums as are understood to
exist in the United States, for the reception of those incorrigible
drunkards, in whom the power of self-control has been altogether
destroyed by their repeated yielding to the craving for Alcoholic
stimulants.

36. *Mental Debility in the Offspring.* — It is scarcely necessary
to accumulate further proof in support of the assertion, that, of all
the single causes of Insanity, habitual Intemperance is the most
potent, and that it aggravates the operation of other causes.   We
have now to show that it has a special tendency to produce Idiocy,
Insanity, or Mental Debility, *in the offspring.*   Looking to the decided tendency to hereditary predisposition in the ordinary forms
of Insanity; looking also to the fact that any perverted or imperfect
conditions of the nutritive functions established in the parent, are
also liable to manifest themselves in the offspring, (as shown in the
transmission of the gouty and tubercular diatheses [1]); we should
expect to find that the offspring of habitual drunkards would share
with those of lunatics in the predisposition to insanity, and that

---

[1] *Tubercular diathesis* — a condition of the system predisposing to scrofula
or consumption.

they would, moreover, be especially prone to intemperate habits. That such is the case, is within the knowledge of all who have enjoyed extensive opportunities of observation; and the fact has come down to us sanctioned by the experience of antiquity. Thus Plutarch says, "One drunkard begets another;" and Aristotle remarks that "drunken women bring forth children like unto themselves." Dr. W. A. F. Browne, the resident Physician of the Crichton Lunatic Asylum at Dumfries, makes the following statements:—"The drunkard not only injures and enfeebles his own nervous system, but entails mental disease upon his family. His daughters are nervous and hysterical; his sons are weak, wayward, eccentric, and sink insane under the pressure of excitement, of some unforeseen exigency, or of the ordinary calls of duty. At present I have two patients who appear to inherit a tendency to unhealthy action of the brain, from mothers addicted to drinking; and another, an idiot, whose father was a drunkard." [1] The author has learned from Dr. Hutcheson, that the results of his observations are precisely in accordance with the foregoing.—On this point, however, the most striking fact that the writer has met with, is contained in the Report on Idiocy lately made by Dr. Howe to the legislature of Massachusetts. "The habits of the parents of 300 of the Idiots were learned; and 145, or nearly *one-half*, are reported as 'known to be habitual drunkards.' Such parents, it is affirmed, give a weak and lax constitution to their children; who are, consequently, 'deficient in bodily and vital energy,' and predisposed by their very organisation to have cravings for alcoholic stimulants; many of these children are feeble and live irregularly. Having a lower vitality, they feel the want of some stimulation. If they pursue the course of their fathers, which they have more temptation to follow, and less power to avoid, than the children of the temperate, they add to their hereditary weakness, and increase the tendency to idiocy in their constitution; and this they leave to their children after them. The parents of case No. 62 were drunkards, and had seven idiotic children." [2]

37. There is a prevalent impression that idiocy is particularly liable to occur in the offspring of a procreation that has taken place when one or both of the parents were in a state of intoxication. A striking example of this kind is related in the *Phrenological Journal* (vol. vii. p. 471); both the parents were healthy and intelligent, and one at least habitually sober; but both were partially intoxicated at the time of the intercourse, and the offspring was completely idiotic. There is every reason to believe that the monomania of inebriety not only acts upon, and renders more deleterious, whatever

---

[1] Moral Statistics of Glasgow, by William Logan, 1849, p. 20.
[2] American Journal of Medical Sciences, April, 1849, p. 437.

5

latent taint may exist; but vitiates or impairs the sources of health for several generations. That the effects of drunkenness are highly inimical to a permanent healthy state of the brain, is often proved at a great distance of time from the course of intemperance, and long after the adoption of regular habits.

38. *Inflammatory Diseases of the Brain.*—All medical writers agree in regarding Intemperance as one of the conditions which tends to produce *Inflammatory* diseases of the Encephalon, now distinguished as *Cerebritis*[1] and *Meningitis;*[2] and this is precisely what might be anticipated, when it is considered how great must be the derangement of the circulating and nutritive operations, occasioned by the presence of alcohol in the blood. An attack of acute Encephalitis[3] not unfrequently supervenes upon a debauch, which is then regarded as its exciting cause.[4] But it may occur quite independently of any special act of excess, in consequence of the predisposition arising from the perversion of the normal functions, by the habitual use of alcoholic liquors in quantities that may never produce actual intoxication. Perhaps, indeed, this is the more common occurrence. We have seen that the state of excitement first produced in most persons by the ingestion of alcohol, would pass into *Meningitis,* (or rather inflammation of the *convolutions*[5]) if it were not to subside with the elimination of the alcohol from the blood. On the other hand, the state of torpor of the mental functions, which alcohol produces from the first in some individuals, and which comes on in all if the intoxication be carried far enough, is indicative of that congestion of the *substance* of the brain, which, if confirmed, and accompanied by a certain disturbance of the nutritive operations, would become *Cerebritis.*[6] There can be no hesitation, therefore, in admitting the relation of cause and effect, in cases in which it is so obviously established by the sequence of the phenomena.

39. There is another class of diseases of the Brain, which are usually dependent upon structural changes that require a longer period for their development; yet whose frequent connection with habitual intemperance is established both by theory and observation. These are Apoplexy, Paralysis, and Epilepsy.

40. *Apoplexy.*—The state of profound Coma, characteristic of the advanced stage of intoxication, may be considered to be identical with that of congestive apoplexy, in every respect save the nature

---

[1] *Cerebritis*—inflammation of the brain.
[2] *Meningitis*—inflammation of the membranes of the brain.
[3] *Encephalitis*—inflammation of the brain.
[4] See, for example, a fatal case related by Dr. Percy, *Op. cit.,* p. 54.
[5] *Convolutions*—the surface of the brain, which has an appearance of being convoluted or thrown into folds.
[6] *Cerebritis*—inflammation of the brain.

ot its cause, and its duration.[1] A certain degree of tendency to
Apoplexy may be said to exist in the slighter form of intoxication;
the vessels of the Brain being congested, as a consequence of increased action of the heart, and of obstruction to the encephalic
circulation, such as is occasioned by imperfect discharge of the functions of the brain; and this obstruction being also favoured by that
partial stagnation of blood in the lungs, which takes place whenever
the respiratory movements are interfered with. This apoplectic
tendency seems to render the intoxicated man peculiarly liable to
suffer from causes, which would not otherwise produce rupture of
the vessels; thus, there are numerous instances on record, in which
blows received in pugilistic encounters, or other comparatively slight
injuries, have occasioned fatal hemorrhage within the cranium;[2] the
sufferer having been previously dosed with spirits in such quantity
as of itself to produce a state of congestion bordering on apoplexy.
And it occasionally happens, though this is comparatively rare, that
cerebral hemorrhage occurs without any external violence, after an
excessive indulgence in spirituous potations.

41. But the influence of Alcoholic Liquors in the causation of
Apoplexy is usually of a much more gradual nature. A large proportion of the cases of Apoplexy occurring in plethoric[3] subjects,
and not connected with disease of the heart or softening of the
arterial coats, are traceable to intemperance in eating as well as in
drinking; the latter, however, being the chief cause, inasmuch as,
without the habitual assistance of alcoholic liquors, continual excess·
in eating would generally soon correct itself. Hence we find that
such cases are rather apt to occur among those who take considerable
quantities of wine or malt-liquor with full meals of solid food, than
among the drinkers of spirits, who are seldom great eaters. It is
not difficult to see the reason of this. For, on the one hand, the
habit of excess in eating and drinking has a tendency to produce
that condition of Plethora,[4] which is most peculiarly prone to favour
hemorrhagic effusions; whilst, on the other, the ingestion of a large
quantity of solid food, by causing pressure on the vessels of the
abdominal viscera,[5] and by impeding the descent of the diaphragm,[6]

---

[1] Although, as we have already seen, (§ 14,) the phenomena are so nearly
identical, the difference in the etiology involves an important difference in
the treatment; the comatose drunkard not requiring, nor bearing, the free
depletion that is proper in a case of true congestive apoplexy.
[2] *Hemorrhage within the cranium* — an escape of blood from the vessels
within the skull.
[3] *Plethoric* — full of blood; full habit.
[4] *Plethora* — fullness of blood.
[5] *Abdominal viscera* — the stomach, intestines, liver, &c.; the organs contained within the belly.
[6] *Diaphragm* — a muscular partition which divides the cavity of the chest
from that of the belly.

tends to force an unusual quantity of blood into the encephalic vessels, as well as to obstruct its return from them. Such an habitual derangement of the circulation may well be supposed to occasion a progressive weakening of the vessels of the brain; and in this manner it happens that after a persistence for months or years in this course, Apoplexy may supervene, and be its legitimate consequence, without the attack being traceable to any extraordinary indulgence.[1]

42. Of the strength of the general opinion of the Medical Profession,—as to the tendency of Alcoholic stimulants to produce the sthenic[2] form of Apoplexy, it is impossible to give a stronger proof than the rigidity of the rule of abstinence which is laid down for those, in whom a disposition to it has already manifested itself. Now if it be necessary to lay down such rules to prevent the recurrence of the disease, is it not most obvious that we are justified in attributing to an habitual violation of them its first occurrence? And if habitual excess be so obviously a predisposing cause, can we reasonably deny that the long-continued even "moderate" use of stimulants is likely to exert a slow, but in the end a decided influence? It is surely in vain here to reply, that as *food* is wholesome in moderation, but is hurtful in excess, so may alcohol be also; for alcohol (as will be more fully shown hereafter) can never properly act as food, save when other alimentary matters are deficient; and even in the smallest and most diluted doses, alcohol exerts an influence on the vital properties of the tissues with which it is brought into contact, that is never manifested by proper alimentary matters.

43. *Paralysis*[3] *and Epilepsy*[4]. — As the conditions upon which the cerebral forms of *Paralysis* depend, are so nearly the same with those which induce Apoplexy, we cannot doubt that the continual intemperate use of Alcoholic liquors must predispose to this disease, especially when it accompanies intemperance in eating; and should expect, too, that an attack of it may sometimes be traced to some particular excess, as its exciting cause. All medical writers accord in stating that such is the result of actual observation; and here, again, we find in the rules of treatment laid down, an additional evidence of the general conviction of the tendency of alcoholic liquors, even in small quantities, to induce a recurrence of paralytic attacks. The writer has had opportunities of noticing

---

[1] There is evidence that habitually excessive use of Alcoholic liquors has a tendency to produce Hemorrhages *elsewhere*, probably by diminishing the plasticity of the blood, and by impairing the nutrition of the walls of the blood-vessels.—(See §§ 52 and 68.)

[2] *Sthenic* — attended with strength and not debility.

[3] *Paralysis* — palsy.

[4] *Epilepsy* — a convulsive disease popularly known as the falling sickness.

this, in the case of two gentlemen advanced in life, each of whom suffered from repeated attacks of paralysis, which almost invariably supervened upon a violation of the habitual rule of abstinence from fermented liquors and of extreme moderation in diet.—Precisely the same, too, may be said of *Epilepsy,* which disease is now generally attributed to a disordered state of *nutrition* of the brain, of which the paroxysm is the manifestation. Of this disordered state of nutrition, intemperance in eating and drinking is among the most frequent of the predisposing causes, especially when the disease occurs in persons advanced in life; whilst in those who are already predisposed from these or other causes, the excessive use of fermented liquors is frequently the immediate or exciting cause of the paroxysm.

44. Besides these positive diseases, a premature exhaustion of Nervous power, manifested in the decline of mental vigour and of nervo-muscular energy, are ranked by common consent among the consequences of habitual excess in the use of Alcoholic liquors; and reasons will be given hereafter for the belief, that it is occasionally the direct, but more frequently the indirect consequence of the habitual employment of what is considered a very moderate allowance.—(See §§ 177, 178.)

45. In regard to all the forms of Encephalic disorder which result from the long-continued action of causes that impair its nutrition, it is to be observed that the habitual use of Alcoholic liquors has,—in addition to its *direct* action upon the functions of circulation and nutrition,—an important *indirect* agency; inasmuch as, by the temporary support it affords, it sustains the nervous apparatus under a degree of exertion that is in the end most injurious to it, and renders the whole system more tolerant of morbific causes of various kinds; the manifestation of whose action, however, is only postponed, and becomes more severe in the end, in proportion to the duration of the agency. This indirect operation of alcoholic liquors, however, will be more fitly considered at a future period.—(§ 198.)

Diseases of the Alimentary Canal.

46. The disorders of the Nervous system, whose symptoms are among the most obvious and characteristic results of Alcoholic Intoxication, having been now considered, we proceed to examine the influence of Alcoholic liquors on the production of diseases of the Digestive Apparatus. This influence is exerted in two ways;—first, by the direct irritating action of the fluid upon the mucous lining[1] of the Alimentary Canal[2]; and second, by the general de-

---

[1] *Mucous lining* — the innermost coat.
[2] *Alimentary canal* — the stomach and intestines.

terioration of the nutritive processes, resulting in various ways from
the entrance of Alcohol into Blood.

47. *Irritation and Inflammation of the Mucous Membrane of
the Stomach.* — That irritation would be produced in the very vascular mucous membrane of the Stomach, by the direct contact of
Alcoholic liquors, and that this would vary in its intensity with the
amount, concentration, and duration of the application of the irritant, is precisely what we should anticipate, from what has been
already shown by observation to be the result of the application of
alcohol to a living membrane. ' A small quantity of alcoholic liquor,
diluted by the fluids already in the stomach, appears to produce only
the first effect, namely, a quickening of the circulation, and a temporary exaltation of the functional activity of the organ, as shown
in the increase of appetite and of digestive power. But when a
larger quantity is introduced, and especially when successive doses
are taken so as to keep up the irritation, or when the alcohol is in
a state of high concentration, and the stomach contains but little
other fluid, all the effects of an irritant are produced, varying from
moderate congestion[1] with diminished functional activity, to intense
congestion passing into inflammation, and even to a gangrenous
state. The more severe effects, however, are not often seen; in
consequence, it may be surmised, of the rapidity with which the
alcohol has been absorbed, (§ 18), and the brevity of the duration
of its contact with the membrane, shielded as this is with its coat
of mucus. Hence a *repetition* of the dose seems more likely to
produce a state of high irritation, or of inflammation, than any
single dose, unless this have been too great to be quickly absorbed.

48. The morbid appearances found in the Stomachs of men or
animals killed by narcotic[2] poisoning, and attributable at first sight
to the direct influence of the irritant, can seldom be fairly regarded
in that light; since they are for the most part such as are producible
by the Asphyxia, which has been the immediate cause of death.
When we find general injection of the mucous membrane, local
patches of extreme congestion, numerous minute extravasations,[3] or
hemorrhagic patches[4] of large extent, these are more likely to have
been the result of the stagnation of the pulmonary circulation, acting backwards upon the whole venous system, than to have been
the immediate result of the contact of alcohol; since appearances
precisely similar are found when death has taken place from suffocation in other modes,—e. g. in Criminals executed by hanging. In

---

[1] *Congestion* — too great fullness of the blood-vessels.

[2] *Narcotic* — having the power to diminish sensibility and consciousness
—opium may be taken as the type of the narcotic poisons.

[3] *Extravasations* — spots where blood has escaped from the vessels beneath or within the substance of the membrane.

[4] *Hemorrhagic patches* — similar extravasations of greater extent.

the case of animals poisoned by Alcohol, it frequently happens that scarcely any positively morbid appearances are discernible in the stomach; and the departures from the healthy character which are noticed in the stomach of the human subject after death from alcoholic poisoning, are most frequently such as indicate an altered state of its nutrition, consequent upon habitual irritation. Of these departures, a *thickened* state of the mucous membrane seems to be the most constant; the membrane being sometimes softened (as stated by Dr. Ogston); sometimes unusually firm, corrugated,[1] and pale (as observed by Dr. Peters). These last appearances seem to have been most common, when a quantity of undiluted Spirits had been taken shortly before death, and to have resulted from that *physical* action exerted by them upon the membrane, to which reference has already been made (§ 2, 3). It sometimes happens, however, that after the narcotic effects of the Alcohol have passed off, another set of symptoms appears, indicative of inflammation of the Alimentary Canal; and if these proceed to a fatal termination (as now and then occurs), the usual appearances indicative of that state are found in the gastro-intestinal mucous membrane.[2] In one example of this kind, cited by Dr. Christison, the whole villous coat[3] of the stomach was in a gangrenous state, the colon[4] was much inflamed, and the small intestines red along their whole length.

49. Our best information as to the effect of Alcoholic liquors upon the condition of the Gastric mucous membrane during life, is derived from the well-known observations of Dr. Beaumont in the case of Alexis St. Martin. This man appears to have been habitually temperate and healthy; but to have occasionally indulged in excess both in eating and drinking, the results of which could be seen by direct observation through the fistulous opening in the parietes of his stomach. Thus, says Dr. Beaumont, under the date July 28th, 1833,—"Stomach not healthy, some erythema,[5] and aphthous[6] patches on the mucous surface. St. Martin has been drinking ardent spirits pretty freely, for eight or ten days past,—complains of no pain, nor shows symptoms of general indisposition,—says he feels well, and has a good appetite. August 1st,—Inner membrane of the Stomach morbid; considerable erythema, and some aphthous patches on the exposed surface; secretions vitiated. August 3rd,—Inner membrane of Stomach unusually morbid; the

---

[1] *Corrugated* — thrown into folds, puckered.
[2] *Gastro-intestinal mucous membrane* — the lining membrane of the stomach and intestines.
[3] *Villous coat* — the innermost or lining membrane; its surface being supposed to resemble velvet.
[4] *Colon* — the largest of the intestines.
[5] *Erythema* — superficial inflammation like erysipelas.
[6] *Aphthous patches* — small ulcers, covered with a whitish matter.

erythematous appearance more extensive, and spots more livid than usual, from the surface of which exuded small drops of grumous[1] blood; the aphthous patches larger and more numerous; the mucous covering thicker than common, and the secretions much more vitiated. The gastric fluids extracted this morning were mixed with a large proportion of thick, ropy mucus, and considerable muco-purulent[2] matter, slightly tinged with blood, resembling the discharge from the bowels in some cases of chronic dysentery." Now, it is very important to remark, that all this disorder was proved by direct observation to be actually existing in the Mucous coat of the stomach, without any such manifestation of it by general or local symptoms, as would by themselves have been thought indicative of its presence. — "For," continues Dr. Beaumont, "St. Martin complains of no symptoms indicating any general derangement of the system, except an uneasy sensation, and a tenderness at the pit of the stomach, and some vertigo, with dimness and yellowness of vision on stooping down and rising again; has a thin yellowish brown coat on his tongue, and his countenance rather sallow; pulse uniform and regular, appetite good, rests quietly, and sleeps as well as usual." By the 6th of August, the inner surface of the stomach had recovered its healthy appearance; the patient having in the meantime entirely abstained from all alcoholic liquors, and having been confined to low diet. Dr. Beaumont further states that "diseased appearances similar to those mentioned above, have frequently presented themselves in the course of my experiments and observations. They have generally, but not always, succeeded to some appreciable cause. Improper indulgence in eating and drinking has been the most common precursor of these diseased conditions of the stomach. *The free use of ardent Spirits, Wine, Beer, or any intoxicating liquor, when continued for some days, has invariably produced these morbid changes."*

50. From the precise concurrence of these Observations with what Theory would lead us to expect, in regard to the action of Alcoholic liquors on the Mucous membrane of the stomach, it is obvious that we have no right to suppose that the peculiar condition of St. Martin gave him any peculiar liability to suffer in the manner above described. On the contrary, such disorders of the circulation, nutrition, and secretion might be anticipated to occur in every case; and it is only because they are not immediately indicated by pain and heat in the stomach, by loss of appetite, or by general febrile disturbance, that they are presumed not to exist. This presumption, however, has been shown to be altogether fallacious; and we have adequate reason to believe that some such condition must

---

[1] *Grumous* — clotted, thick.
[2] *Muco-purulent* — a mixture of mucus and matter.

be the result of *every excess* in the use of alcoholic liquors, however little it may be indicated by the local or general symptoms.

51. *Inflammatory Gastric Dyspepsia.*—It might be anticipated, then, that habitual excess would convert this state of occasional and transient disorder, which only requires rest and abstinence for its cure, into one of a more persistent and obstinate character; which, by unfitting the stomach for the discharge of its normal functions, would seriously impair the general nutritive operations. Such has been shown by experience to be the case; a special form of dyspeptic disorder, termed *Inflammatory Gastric Dyspepsia*, being well known to practical men, as common among those who have freely indulged in alcoholic potations. Of this disorder, the following are the symptoms, as enumerated by Dr. Todd:[1]—"Painful digestion, sense of heat, tenderness, or pain at the epigastrium,[2] increased upon taking food, or on pressure; thirst; tongue more or less of a bright red colour, sometimes brownish red, sometimes dry, glossy, and adhesive; taste saltish or alkaline, occasionally like that of blood; bowels generally confined; urine high-coloured; skin dry, with occasionally profuse, partial sweats, chiefly in the direction of the extensor muscles; temperature of the trunk increased, of the extremities diminished, except occasionally in the palms of the hands and soles of the feet, which, especially at night, are frequently hot, dry, and burning; aggravation of the symptoms under the use of stimulants or of irritating ingesta."[3] The various stages and degrees of the disease are characterized by various modifications of these symptoms, many of them the consequences of the disturbance of the nutritive functions produced by the disorder of the stomach; but of all such consequences it may be remarked, that they are probably aggravated by the previous disturbance of the nutritive and secretory operations, consequent upon the habitual introduction of alcohol into the blood. Thus we find a special tendency to cutaneous eruptions, such as Erysipelas,[4] Lichen,[5] Erythema,[6] Urticaria,[7] Psoriasis,[8] and Pityriasis;[9] to sluggish and imperfect action of the Liver; to scantiness in the secretion of the Kidneys; and to depression of spirits, with inability for active mental exertion, passing on, in the more confirmed states, to complete Hypochondriasis. Although excess in eating may aid in the production of this

---

[1] Cyclopædia of Practical Medicine, Art. *Indigestion.*
[2] *Epigastrium* — the region of the stomach.
[3] *Ingesta* — substances taken into the stomach.
[4] *Erysipelas* — inflammation of the skin.
[5] *Lichen* — an eruption upon the skin of red pimples.
[6] *Erythema* — superficial inflammation.
[7] *Urticaria* — the nettle rash.
[8] *Psoriasis* — a disease of the skin attended with patches of rough scales.
[9] *Pityriasis* — a disease of the skin characterized by irregular patches of thin scales.

wretched condition, yet, as Dr. Todd remarks, it is rather due to the stimulating quality of what is taken into the stomach, than to its quantity; and although it may occasionally arise from the habitual use of highly-seasoned food without the proper dilution by bland liquids, yet it is much more frequently brought on by indulgence in alcoholic potations; " it is the dyspepsia of the Dram Drinker and Opium Eater, and belongs altogether more to the Drunkard than to the Glutton." In the treatment of this disease, the complete disuse of stimulants is found to be of the greatest importance ; notwithstanding that, in the more chronic forms of it, a temporary alleviation is sometimes obtained from small quantities of alcoholic liquors.[1]

52. *Disorders of the Intestinal Mucous Membrane.*—The disordered state of the gastro-intestinal mucous membrane[2] is not limited, as we have seen, to the stomach, and it may extend itself along the whole course of the Alimentary Canal, to parts with which the Alcoholic liquors themselves have not come in contact; so as to be attributable rather to the general imperfection of the nutritive operations, than to the local effects of the stimulant. Thus we find that habitually intemperate persons are subject to soreness, redness, and ulceration of the membrane of the nose, and of that of the lower part of the intestinal canal; and *hemorrhages*[3] from various parts of this membrane, as well as from the mouth itself, are of no unfrequent occurrence,—the escape of blood being obviously dependent in part on its own insufficient plasticity,[4] and in part upon the softened condition of the walls of the vessels. It is important to bear this in mind, as increasing the probability of the same cause being concerned in the production of a similar softening elsewhere; as, for example, in the vessels of the Brain.—(See §41).

53. Where, in place of excessive indulgence, what is commonly considered a *moderate* use has been made of Alcoholic liquors, we cannot with the same confidence attribute to it any decided departure from the healthy condition of the Stomach ; and it is certain that the mucous membrane becomes in time so habituated to its presence, that its contact no longer produces the same effects as it does on a membrane unaccustomed to it. But we shall hereafter (§ 160–162) find reason to believe, that such habitual use is not without its consequences, although these may be very remote; the continual over-excitement of the vital activity of the gastric mucous membrane being probably one of the causes of that premature loss

---

[1] See the observations of Sir Philip Crampton on this subject, in Dublin Hospital Reports, vol. i. p. 349.

[2] *Gastro-intestinal mucous membrane* — inner or lining membrane of the stomach and intestines.

[3] *Hemorrhages* — discharges of blood.

[4] *Plasticity* — consistence, capacity to become organized.

of functional power, which is observable in a great number of those who have accustomed themselves to the use of alcoholic liquors. This cause, however, will seldom act alone; being usually combined with excess in diet, and with "wear and tear" of the general system, as will be shown in its proper place; so that its operation is very liable to be overlooked.

Diseases of the Liver.

54. That habitual excess in the use of Alcoholic liquors must have a direct tendency to produce certain diseases of the Liver, will be questioned by no one who considers their mode of introduction into the system, and their influence on the condition of the blood. The blood which returns from the gastric veins charged with Alcohol, is immediately transmitted through the Liver; and it stimulates this gland for a time to increased activity, one effect of which is to eliminate a portion of the alcohol from the blood,—this substance, according to Dr. Percy's observations, being detectible in the bile of animals poisoned by alcohol. Hence the Liver, like the stomach, is subject to habitual over-stimulation from the direct contact of alcohol with its substance. But we have seen that the presence of alcohol in the blood prevents it from acquiring its proper arterial character by passage through the lungs; and we shall hereafter find that it causes the undue retention in it of hydro-carbonaceous matters,[1] which ought to be removed by the respiratory process. Hence an undue amount of labour is thrown upon the Liver,—one of the functions of this Gland being, to separate from the blood such hydro-carbonaceous matters as are not carried off by the respiratory organs; and this continual over-work must predispose it to various disorders.

55. *Acute and Chronic Inflammation of the Liver.* — In tropical climates *acute inflammatory diseases* of the Liver are among the most common of these disorders; and they are distinctly traceable, in a large proportion of cases, to that excess both in eating and drinking, to which Europeans are unfortunately but too prone; being rare among the Natives, and almost equally rare among the Europeans who adopt the native manner of living. In this country such acute diseases are comparatively rare; but there are certain remote consequences which are no less clearly traceable to *chronic inflammation and degeneration*, resulting from the excessive use of fermented liquors, especially when these are taken in the form of distilled spirits.—The following is the account of the state of the Liver given by Dr. Peters (*loc. cit.*), as presented in the seventy cases which he had an opportunity of examining. "In 'moderate drinkers,' the liver was generally found to be somewhat larger than

---

[1] *Hydro-carbonaceous matters* — substances composed of hydrogen and carbon.

usual, its texture softened, and its outer surface spotted, with patches of fatty infiltration extending two or three lines into the parenchymatous [1] substance; the rest of the viscus retaining its natural colour, and its edges their normal sharpness. In those who had been more addicted to the use of spirits, the liver was still larger, its edges were more obtuse, and the patches of fat on its surface were larger and more numerous. In old drunkards the liver was very large, weighing at least six .or eight pounds, often from ten to twelve; the edges were very thick and much rounded; the parenchyma [2] almost white with fat, soft, fragile, and the peritoneal covering could be torn off with ease." It is evident that in all these cases, the Liver was the subject of various degrees of *fatty degeneration*, which takes place, on the one hand, as the result of deficient functional activity of the Gland, whilst on the other it is indicative of an excess of fatty matter in the system.

56. The peculiar conditions of the Liver known as " granular liver," and " hob-nailed.liver," or " gin liver," were comparatively rare in Dr. Peters's experience, being observed only in four or five cases; but they seem to be much more common in this country; and its greater prevalence may possibly be due to a difference in the character of the spirit usually employed by drinkers among the lower classes, *Gin* being here the most common, *Rum* and *Brandy* in the United States. These conditions appear to be dependent upon *atrophy* [3] of the proper hepatic substance, with *hypertrophy* [4] of the connecting areolar tissue; the former being apparently the result of the exhaustion of the functional power of the liver by over-excitement; and the latter to continual attacks of chronic inflammation, which produce the false membranes, adhesions, puckerings, &c., that give rise to the second of the designations just cited. Between the state of *contraction* (in which the Liver is frequently not more than half its usual size), and the state of *enlargement* just described, there is not that opposition which might at first sight appear; for in both is there diminished functional and nutritive activity of the proper substance of the gland; and the state of enlargement, which is simply dependent upon the accumulation of fatty matter, not unfrequently gives place to one of contraction. In fact, it would not seem improbable that each state may have a relation to the general disposition to the development of fat, in the individual; for whilst ir. many habitual drunkards there is a great tendency to the production of fat, and to its deposition in various parts of the body (§ 61), there is an equal tendency in others to a leanness which no fatten-

---

[1] *Parenchymatous substance* — the substance or tissue of the gland.
[2] *Parenchyma* — see the preceding note.
[3] *Atrophy* — a wasting away, diminution of bulk.
   Hypertrophy — excessive growth, augmentation of bulk.

ing process will overcome.—Certain it is, however, that the habitual use of Alcoholic liquors has a tendency first to excite and then to diminish the functional activity of the Liver; and thus predisposes in the first instance to inflammatory diseases of the organ, whilst its more remote operation is to induce atrophy or degeneration. This will be especially the case in tropical climates; where several causes concur (as will be shown hereafter) to augment the injurious influence of Alcohol upon the Liver, and consequently to increase the amount and severity of the diseases of that organ induced by its habitual use.—Of course, every disturbance of the function of the Liver must be an additional source of disorder in the digestive operations, in which the action of this gland has so important a share.

Diseases of the Kidneys.

57. We have seen that a special determination of blood to the Kidneys takes place as one of the results of the reception of Alcoholic liquors into the blood; and these organs are thereby excited to augmented action, one of the purposes of which would seem to be the removal of the alcohol from the current of the circulation. As the blood of the Kidneys is derived from the arterial system, in which the alcohol becomes diluted by the whole mass of sanguineous fluid; and as the alteration in the constituents of the blood which it tends to produce, has less relation to the function of the kidneys than to that of the liver; it might be expected that excess in alcoholic liquors should not have the same tendency to produce acute inflammatory attacks in this organ as in the other, although it may act as the exciting cause of such attacks (as appears to be frequently the case), when the predisposition has been established by other agencies. But we should expect that the habitual use of alcoholic liquors in excess would have a special tendency to produce a state of *chronic irritation*, passing into chronic inflammation, with various consequent alterations in the structure, and deterioration in the function, of the Kidneys.—Such we have every reason to believe to be the usual origin of that morbid condition commonly known as *Bright's Disease*, or *Granular Degeneration* of the Kidneys, which is now generally considered by Pathologists as a result of chronic inflammation and atrophy of the proper substance of the Kidney, with deposits of fatty, albuminous, or other unorganizable matters,—a state, in fact, very closely resembling the degenerated conditions of the Liver already described. Now of this disease Dr. Christison states that from three-fourths to four-fifths of the cases which he met with in Edinburgh, were in persons who were habitual drunkards, or who, without deserving this appellation, were in the constant habit of using ardent spirits several times in the course of the day; and the experience of English Hospital practice is (so far as the writer has

6

been able to ascertain) precisely similar. The disease is very rarely met with in the private practice of those, whose patients are of a class not given to excessive spirituous potations. Here, too, it would seem as if the use of malt spirit (Gin or Whiskey) gives a greater predisposition to the disease, than that of Rum or Brandy; the former having a more diuretic effect than the latter, that is, producing a greater temporary activity in the kidneys, and having a greater tendency to bring about a state of chronic irritation.

58. But we are not to suppose that, if this severe form of renal disease be not developed, the Kidneys escape altogether free. We should expect that the consequences of long-continued and habitual excitement would manifest themselves in subsequent impairment of functional power, even if no obvious structural disease be engendered; and there can be little doubt that such is the case, since we find that persons advanced in life, who have habitually indulged freely, even if not excessively, in Alcoholic liquors, are extremely apt to suffer from *Gout, Rheumatism,* and other disorders, which mainly depend upon the insufficient elimination of such morbid matters from the blood, as ought to be carried forth through this channel (§ 66). Excesses in diet, which, at an early period of life, are counteracted by the activity of the excretory apparatus, are no longer thus prevented from giving rise to an accumulation of morbific[1] products in the blood, when the Kidneys begin to fail in the performance of their duty; and although we may not be able with positive certainty to attribute this failure to free indulgence in alcoholic liquors, yet it cannot be reasonably questioned that such habits must tend to produce it,—since we find that overexcitement of *any* organ is regularly followed, sooner or later, by depression of its functional power, and have seen that the continual stimulation of the Kidney by alcohol has a special tendency to produce perverted nutrition, and thus to render it entirely unfit for the performance of its duties.

Diseases of the Skin.

59. The determination of blood to the Skin, which has been noticed as one of the results of the ingestion of Alcoholic liquors, has a tendency, when frequently repeated, to produce various disorders in its nutrition, chiefly those resulting from congestion or inflammation of its several tissues. Such disorders show themselves especially in the skin of the face; and this for two reasons,—because, in the first place, the face partakes in the general determination of blood towards the head, so that it becomes more flushed than any other part of the surface; and also because the exposure of this part of the cutaneous surface disposes it to be more affected

---

[1] *Morbific* — disease-producing.

than that of the body and limbs by external cold, which will al-. ways tend, by lowering the vital activity of any tissue, to increase the evils resulting from too copious a determination of blood towards it. Hence we find the skin of the face especially disposed to exhibit those Carbuncles, Boils, &c., which may be considered, in a large proportion of cases, as the direct result of habitual intemperance; it is also the part in which the Erysipelatous attacks, so common among the intemperate, most frequently commence, when · they are not immediately excited by some injury elsewhere; and it is on the face, too, that we most frequently meet with various forms of Acne, of which the *Acne rosacea*[1] is, in a very large proportion of cases, directly attributable to intemperate habits. We have already noticed other diseases of the Skin (§ 51), which seem to be rather consequent upon the disorder of the digestive apparatus induced by the habitual free use of Alcoholic liquors, than due to the direct agency of the alcohol upon its tissue. There is a disease, however, noticed by Dr. Darwin under the name of *Psora Ebriorum*,[2] which may be attributed with great probability to a chronic though slight perversion of the nutritive operations of the skin, in consequence of the presence of alcohol in the blood. Of this disease Dr. Darwin says: "Elderly people who have been much addicted to spirituous drinks, as beer, wine, or alcohol, are liable to an 'eruption all over their bodies, which is attended with very afflicting itching, and which they probably propagate from one part of their bodies to another, with their own nails, by scratching themselves." Dr. Macnish states that he has himself seen many cases of this disease.[3] Most other cutaneous[4] disorders, which are less directly traceable to intemperate habits, are greatly aggravated by them; so that strict abstinence from fermented liquors is an almost invariable rule in the treatment of them, unless the use of these in small quantities should be thought requisite to improve the state of the digestive function.

General Disorders of Nutrition.

60. Having thus considered the principal forms of disease which the intemperate employment of Alcoholic liquor has a tendency to induce in the several parts of the Excretory apparatus, to which they seem to give a special determination, we have now to consider those General Disorders of Nutrition, which are traceable to the same cause, which manifest themselves either as substantive diseases, as modifying the course of other diseases, or as giving a special liability to the action of other morbific causes. —We have

---

[1] *Acne rosacea* — red pimples or tubercles upon the nose and face.
[2] *Psora ebriorum* — the drunkard's itch.
[3] Anatomy of Drunkenness, p. 178.
[4] *Cutaneous* — appertaining to the skin.

already spoken of the deteriorating effect of the admixture of Alcohol with the Blood; how it lowers the plasticity of the fibrine,
tends to empty the red corpuscles, and in various ways impedes the
process of aeration; and another less direct but not less important
source of deterioration, is to be found in the imperfect elimination [1]
of the constituents of the Bile and Urine, which must be the consequence of functional inactivity, still more of structural degeneration,[2] of the Liver and Kidneys. Hence it would seem impossible that by such a pabulum the formation of the solid tissues can
be normally sustained; and we should expect to find that the
nutritive processes are not performed with the same energy and
completeness in the habitually intemperate, that they are in the
habitually abstinent. Notwithstanding some appearances to the
contrary, there is abundant evidence that such is the case.
Although a high degree of bodily vigour seems to be exhibited by
certain classes of men, who consume large quantities of fermented
liquors, yet this is extremely deceptive, as the facts to be presently
stated will clearly indicate; and the general result is evidently on
the other side.

61. *Tendency to the Deposition of Fat.* — The immediate effects
of Alcoholic liquors upon the general appearance of the body,
especially as regards the deposition of fat, vary with their nature,
and with the circumstances under which they are habitually used.
Thus it is generally to be noticed that those who indulge largely in
malt liquors become fat, and often exceedingly corpulent; the large
consumers of wine commonly share the same tendency; but the
spirit-drinker is more commonly lean and even emaciated. This
difference may partly depend upon the constitution of the liquors;
thus ale, beer, &c., contain a considerable amount of saccharine [3]
matter, which is either consumed in respiration, leaving the fatty
matters of the blood to be deposited as fat, or is itself converted
into fat; in wine, again, there is more or less of solid matter,
which furnishes materials for combustion; whilst in distilled spirits,
there is scarcely anything save the alcohol. But it also depends in
part upon the amount of solid food habitually taken with the
drink; thus the beer-drinker, if he be leading a life of great
muscular exertion, may find his appetite but little impaired by his
excess; the wine-drinker also usually feeds high; whilst the spirit-
drinker, especially among the poorer classes, takes his dram *instead
of* solid food, for which he has neither appetite nor pecuniary
means. The corpulence of the beer and wine-drinker, however,
seldom continues to old age; and the parts which first begin to
shrink are the legs, after which the shoulders generally give way,

---

[1] *Elimination* — separation, removal.
[2] *Structural degeneration* — diseased state of the substance of the organ.
[3] *Saccharine* — having the sweetness and characters of sugar.

and the whole body becomes loose, flabby, and inelastic, the abdomen alone retaining its protuberance, in consequence of the large deposition of fat in the omentum,[1] which is rarely absorbed. Such a deposition of fat is almost invariably found in the omentum of confirmed spirit-drinkers,[2] notwithstanding its absence elsewhere.

62. A general corpulence of the body, however, can by no means be admitted as an indication of healthy nutrition; indeed it must be regarded as very much the reverse. No animal in a state of. nature exhibits any considerable deposit of fat, except for some special purpose (as in the case of Cetacea[3] and other warm-blooded animals inhabiting the water, where the coating of fat serves as a non-conductor; or in the case of hybernating mammals,[4] as also of many birds, whose autumnal accumulation of fat is destined to make up for the deprivation or deficiency of food in the winter): and when by a change of habits the deposition of fat is artificially promoted, it is obvious that the muscular vigour and general "hardiness" of the system are much impaired, the animal becoming liable to many disorders from which it was previously exempt, and requiring much more careful treatment to keep it in good condition. When, indeed, we find a tendency to the deposition of fat, not in *addition to*, but *instead of*, the normal tissues, the case is one of "fatty degeneration," and must be regarded as a positive disease, —involving as it does, a general functionary inactivity.[5]

63. *Diminished Power of Sustaining Injuries by Disease or Accident.*—The classes of men among whom there is an appearance of remarkable bodily vigour, notwithstanding habitual excess in the use of Alcoholic liquors, are those who are continually undergoing great muscular exertion, and who not only drink largely, but eat heartily. Of this class, the London Coal-heavers, Ballasters, and Brewers' Draymen are remarkable examples; many of them drink from two to three gallons of porter daily, and even spirits besides,

---

[1] *Omentum*—a fatty membrane covering the bowels in front.
[2] Dr. Peters, loc. cit.
[3] *Cetacea*—the whale kind.
[4] *Hybernating* — animals that sleep or become torpid during the winter. *Mammals*—animals that suckle their young.
[5] The following interesting case is recorded by Dr. Robertson (Treatise on Diet, 4th Edition, vol. i. p. 272). The subject of it was a very young man, who died thus early from the intemperate use of spirits. For several months before his death, he had been unable to eat more than a very small quantity of food, and his powers were almost exclusively maintained by frequent dram-drinking. The immediate cause of death was cerebral "ramollissement (*softening*):" but although the body was much attenuated, the muscular fibre of the system much wasted, and the sub-cutaneous fat of the extremities had almost disappeared, on cutting through the abdominal walls to examine the condition of the liver, at least three times the usual thickness of fat had to be divided.

6 *

they are for the most part large, gross, unwieldy men, and are capable of great bodily exertion;—so long, at least, as their labour is carried on in the open air.[1] But it does not hence follow that they are in a condition of real vigour; for the constitutions of such men break down before they are far advanced in years, even if they do not earlier fall victims (as a large proportion of them do) to the results of disease or injury which were at first apparently of the most trifling character. It is well known to those who have observed the practice of the London Hospitals, that when such men suffer from inflammatory attacks, or from local injuries, these are peculiarly disposed to run on to a fatal termination; in consequence, it is evident, of the deficient plasticity[2] of the blood, of the low assimilative power of the solids, and of the general depression of the whole vital energy, resulting from habitual over-excitement. The want of plasticity of the blood gives to the inflammatory processes an *asthenic*[3] instead of a *sthenic*[4] character; there is no limitation by plastic effusion,[5] but they spread far and wide through the tissues; depletion cannot be borne; and the only hope of success lies in the use of opium and stimulants with nutritious diet, to sustain so far as possible the prostrated energy. Thus we see that in such men the slightest scratch or bruise will not unfrequently give rise to a fatal attack of Erysipelas; and that internal organs affected with inflammation rapidly become infiltrated with pus,[6] or pass into a gangrenous[7] state. Hence the Surgeon is very unwilling to perform severe operations upon them, knowing that their chance of recovery is but small.—The condition of these men in regard to recovery from injuries, is in remarkable ˆcontrast to that of men who have been "trained" to pugilistic encounters; the latter having been brought to a condition of the highest possible health, by active exercise, abundance of nutritious food, occasional mild purgation, and either entire abstinence from fermented liquors, or by the very sparing use of them. Men thus "trained" recover with remarkable rapidity from the severe bruises which they are liable to receive.

64. Although there are now few men who habitually take *wine* to a corresponding extent, or who maintain by active exercise in the open air any thing like the same muscular vigour, yet such examples are occasionally met with among the fox-hunting country

---

[1] See Appendix A.

[2] *Plasticity* — adhesiveness, disposition to become organized.

[3] *Asthenic* — wanting strength; attended with debility.

[4] *Sthenic* — vigorous; unattended with debility.

[5] *Plastic effusion* — a separation from the blood of an adhesive substance, which, by uniting together the parts surrounding the inflammation, prevents its spreading.

[6] *Pus* --matter produced by ulceration or in a certain stage of inflammation.

[7] *Gangrenous* — mortified; dead and decomposed.

squires, who spend their whole days on horseback, and pass their evenings in drinking port wine. Of these, also, the same remark may be made; that notwithstanding their appearance of vigour, they are bad subjects for medical or surgical treatment, owing to the imperfect condition of their nutritive functions.[1] Among the spirit-drinkers of our large towns, it is notorious that the nutritive and reparative powers are low; and of this fact we have a remarkable illustration, in the frequency, among the intemperate, of a certain form of phagedenic[2] ulceration, whose origin is sufficiently indicated by the term "Geneva-ulcer," by which it is commonly known at Guy's and other Metropolitan Hospitals. This ulcer, usually commencing on the leg, begins as a red, angry, and painful spot, which passes into an open sore; and this increases rapidly, both in depth and breadth, so as even to involve the whole surface of the calf, laying bare the muscles, tendons and nerves. It is not confined, however, to gin-drinkers, but is occasionally met with in the bloated, plethoric, red-faced wine-bibber.

65. *Liability to Epidemic Diseases.* — Another most important indication of the disordered state of Nutrition, consequent upon habitual excess in the use of intoxicating liquors, is the liability of the intemperate to suffer from various other morbific causes,[3] especially those of an Epidemic or Pestilential nature. On this last point, there is, the writer believes, no difference of opinion amongst Medical Practitioners in any part of the world; all being agreed that the habitual drunkard is far more likely to suffer from such agencies, than the habitually sober or temperate man. Whether habitual *abstinence* is still safer than habitual *moderation*, is a point which cannot be so easily ascertained; some considerations on this subject, however, will be offered hereafter (§ 144–150). The peculiar liability of the habitually intemperate to suffer from the Cholera-poison, is well known. The following circumstance, which occurred during the former epidemic of Cholera, is very significant on this point; especially showing that the state of *depression* which follows excitement is the one in which the system is most readily affected. The nurses in the Cholera Hospital at Manchester were at first worked six hours, and allowed to go home the other six; and the mortality was so great amongst them, that there were fears of the failure of the supply. It was found, however, that they were much given to Alcoholic potations (with the idea, probably, of increasing their power of resisting the malady) during their leisure hours; and they were therefore confined to the Hospital, and debarred from obtaining more than a small allowance of alcoholic drink; after which not a

---

[1] *Nutritive functions* — functions concerned in building up the solid parts of the body, and in repairing the waste constantly taking place in them.

[2] *Phagedenic* — spreading, corroding, eating in.

[3] *Morbific causes* — causes producing disease.

single fresh case occurred among them. During the present epidemic, the writer has learned from various sources that a considerable proportion of those, in whom the liability to the disease was not evidently produced by the condition of the locality in which they resided, might be considered as deriving a predisposition to it from habitual Intemperance; — many establishments having lost those men, and those only, who had been accustomed to free indulgence in the use of alcoholic liquors. The general connection between the intemperate habits of a population, and its high rate of mortality from various causes, will be shown hereafter; and a high rate of *mortality* is always indicative of a large amount of *sickness*, although the ratio between the two is by no means constant.

66. *Gout and Rheumatism.* — Among the general disorders of nutrition, to which the intemperate use of Alcoholic liquors certainly predisposes, although it may not of itself cause them, are Gout and Rheumatism. The former is most common among those who have been accustomed both to eat and to drink freely; and it is favoured by such a use of alcoholic liquors, as stimulates the stomach to digest more azotized[1] aliment than the system can appropriate. This may be regarded as the fundamental cause of the disease, when it occurs in its sthenic form. Of the ulterior stages of it, we yet know too little to enable us to trace with certainty the effect of alcohol upon each of them; but this much is pretty certain, — that an impaired condition of the nutritive operations will be favourable to the production of the *materies morbi*,[2] whatever be its nature;—that this will be further promoted by any impediment to the due oxidation of the constituents of the blood, such as the admixture of Alcohol has been shown to occasion; and that the elimination of this morbid matter will be obstructed by that torpid condition of the Liver and Kidneys, to which these organs are especially liable in those who habitually over-excited them in earlier life (§ 58). In the production of Rheumatism, also, we may clearly trace the aggravating influence of habitual excess in the use of alcoholic liquors, especially if the *materies morbi*[2] be, as many suppose, Lactic Acid,[3] or one of its compounds. For whilst the disordered condition of the assimilative and nutritive operations will give a special tendency to the production of this substance, the impediment to its oxygenation[4] presented by the presence of alcohol in the blood will cause it to be retained and to accumulate there, instead of being burned off (which it ought to be, as fast as formed) and escaping from the lungs in the condition of carbonic acid and water.[5] Here, again, the torpor of

---

[1] *Azotized* — containing nitrogen, as animal flesh.
[2] *Materies morbi* — the matter or material causing disease.
[3] *Lactic acid* — an acid obtained from milk.
[4] *Oxygenation* — being supplied with oxygen.
[5] This idea of the influence of alcohol in conducing to the retention of

the Liver and Kidneys, and the disordered action of the Skin, in
the habitually intemperate, will present an additional obstacle to the
proper elimination of the morbific matter; and in Rheumatism, as
in Gout, the intensity of the inflammation can scarcely but be augmented by the diathesis [1] induced by the habitual presence of alcohol in the blood.—All these predictions are verified by the experience of every practical man.

67. *Diseases of the Heart and Arteries.*—Closely connected with
the gouty and rheumatic diatheses are *Diseases of the Heart and
Arteries;* of which some obviously arise out of these constitutional
states, and are thus indirectly favoured by the abuse of Alcoholic
liquors; whilst others seem to be more directly dependent upon the
introduction of alcohol into the blood. The continual but irregular
excitement of the contractile action of the heart and arteries, which
is the result of habitual use of stimulants, must of itself predispose their tissues to disease; and this predisposition will of course
be increased by the contact of blood charged with alcohol with their
lining membrane, as well as by the general disordered condition of
the nutritive operations. Now attacks of acute Arteritis [1] seem not
unfrequently traceable to alcoholic intoxication; and it cannot
therefore, be regarded as improbable, that those more chronic disorders of their walls, which give rise to Aneurism,[2] softening, fatty
degeneration, and other structural changes, and which thereby predispose to hemorrhage, should be favoured, if not absolutely produced, by the habitual presence of alcohol in the circulating
current. Accordingly we find the intemperate use of alcoholic
liquors specified by authors on the Diseases of Arteries, as among
the most important of their predisposing causes.

68. *Spontaneous Combustion.* — Although the phenomenon
termed "Spontaneous Combustion" of the Human body is one of
such rarity, that it might seem scarcely to deserve to be ranked among
the ordinary results of habitual excess in the use of Alcoholic liquors,
yet it should not be passed by in any enquiry into the consequences
of such excess; more especially since it may be regarded with much
probability, as resulting from the same kind of perverted nutrition,
carried to an extreme degree, as that to which we have already
traced various other consequences. It would be more correct to
speak of these cases as instances of *unusual combustibility* of the

---

lactic acid, and thereby favouring the rheumatic diathesis, is confirmed by
the success of Dr. G. O. Rees's method of treating Rheumatism by Lemon
juice; the rationale of which seems to be, that the citric acid affords a
large and ready supply of oxygen, whereby the lactic acid (or *materies
morbi,* whatever it be,) is burned off.

[1] *Diathesis* — condition of body.

[2] *Arteritis* — inflammation of arteries.

[3] *Aneurism* — tumour resulting from the dilatation of an artery in some
part of its course.

body, than of really spontaneous combustion; since in scarcely any of them, perhaps in none, does there seem adequate evidence that the combustion originated without the contact of external flame; their real peculiarity consisting in this — that whereas an ordinary human body requires a large amount of wood, coal, or other inflammable material for its combustion,—the body in the subjects of this accident takes fire very readily, and burns as if it were itself highly inflammable. In some instances it has appeared as if a very inflammable gas were given off from the body; a flame having darted towards it from some distance. In all or nearly all the cases in which the previous habits of the individuals were known, they had been intemperate; and it is remarkable that the greater number of recorded instances occurred among fat old people who had been spirit-drinkers.—The chief source of this peculiar combustibility is probably to be found in the impregnation of the fluids and solids of the body with Phosphorus, which is perhaps united with hydrogen, so as to form some highly inflammable compound. This may be conceived to result from the habitual ingestion of Alcohol, in the following way. The normal mode in which the phosphorus set free by the waste or disintegration [1] of Nervous matter, is extricated from the system, is through the urine, after having been converted by oxidation into phosphoric acid. Now if there be not oxygen enough in the blood to effect this conversion, it is to be expected that the phosphorus would be retained in the fluids, and possibly deposited again in the solids; and since we have seen that the continual presence of alcohol in the circulation gives even to arterial blood a venous character, it is not difficult to understand how much a retention of the phosphorus destined for excretion should be favoured by habitual intemperance. It is a remarkable confirmation of this view, that the *breath* of drunkards has been sometimes observed to be luminous, as if it contained the vapour of phosphorus or of some of its compounds; and that it has been found by experiments upon dogs, that if phosphorus be mixed with oil and injected into the blood-vessels, it escapes unburned from the lungs, if time be not given it to unite with the oxygen of the blood. [2]

The foregoing are the principal disorders, local and constitutional, in the production of which we can trace the operation of the habitually excessive use of Alcoholic stimulants, with tolerable directness. It would be easy to extend this catalogue by the inclusion of other diseases which are manifestly aggravated by intemperate habits; but this, in fact, would require the enumeration of almost

---

[1] *Disintegration* — breaking up; reduction into very minute particles.

[2] See Casper's Wochenschrift, 1849, No. 15. — The luminosity observed by Sir Henry Marsh in the faces of two phthisical patients, was probably due to the same cause — the imperfect oxidation of phosphorus within the body, and its consequent extrication from the skin in a vaporous condition.

every disease to which the human body is subject, more especially if Inflammation participate in it. But the writer thinks it preferable to limit his statements to the cases in which the chain of causation is most continuously and obviously traceable. It has been his object throughout to show what consequences might be expected to arise from habitual "intemperance;" regard being had to the facts which have been fully ascertained, with respect to the *modus operandi* of alcohol on the system at large, and on special organs. It has been shown, that a variety of disorders of the nervous system, of the digestive apparatus, of the secreting organs, of the skin, of the heart and arteries, and of the organic fluids and solids in general, might be thus anticipated; and that such antici pations are all completely verified by the results of practical obser vation.—We might now push the investigation further, and inquir, what evidence we have in regard to the consequences of the habitually "moderate" use of alcoholic liquors on the human system. It must be freely admitted, however, that we have not the same data for the determination of this question, as that on which we have been hitherto engaged; and this on two accounts,—first, that the consequences will be naturally remote, and will be often such as appear fairly attributable, in great part, if not entirely, to other causes;—and second, that the very general prevalence of the "moderate" or "temperate" use of alcoholic liquors, and the shortness of the time during which Total Abstinence has been hitherto practised by any large number of individuals, render it difficult, if not impossible, to draw any valid inference, as yet, from comparative observation. But the writer would argue, that if we have such a complete accordance between the predictions of Theory and the results of Observation, in regard to the consequences of habitual "excess," as establishes the relation of cause and effect beyond dispute; we have a strong case in favour of such a relation, when, the cause being in less active operation, the predicted effects *do* occur, even though at a period so remote as apparently to disconnect them from its influence. For various reasons, however, he deems it advisable to carry out this inquiry under the second head, where it will be more conveniently discussed.

IV. GENERAL EFFECT OF THE EXCESSIVE USE OF ALCOHOLIC LIQUORS ON THE DURATION OF LIFE.

69. We shall close this part of the inquiry by examining into the general tendency of the excessive use of Alcoholic liquors to shorten life; either by themselves giving rise to the diseases above enumerated, or by increasing the susceptibility of the system to other morbific causes. That such a tendency exists, cannot for a moment be questioned. No Life Insurance Office will accept an Insurance on an individual whose habits are known to be intemperate; and

if it be discovered after his death that he has been accustomed to the excessive use of alcoholic liquors, contrary to his statement in his proposal for insurance, the Policy is declared void. And it is, doubtless, owing in part to the superior sobriety of the great bulk of Insurers over that of the average of the population, that a lower rate of mortality presents itself amongst them, than that which might be expected according to the calculations founded on the entire mortality of the country, — to the great profit of the Office. Thus at the age of 40 years, the annual rate of mortality among the whole population of England, is about 13 per 1000; whilst among the lives insured in the Life Offices, it is about 11 per 1000; and in those insured in Friendly Societies, it is about 10 per 1000. —Now the average mortality for *all ages* between 15 and 70 years, is about 20 per 1000; whereas in the Temperance Provident Institution, after an experience of eight years, and with several lives above 70 years of age, the average mortality has been only 6 per 1000, up to the present season, in which it has undergone a slight increase from the Cholera epidemic. It is worthy of remark, however, that although many of the insurers in this Office are of the poorer class, whose condition and employments expose them much more than the middling classes generally to the endemic[1] causes of Cholera, no more than 8 have died of this disease, out of the total of about 3500 insurers. As a means of further comparison, the following table may be subjoined, in which the mortality of the insurers in the Temperance Provident Institution, for the first five years, is compared with that of the insurers in other Offices during the corresponding period of their existence.

LIFE POLICIES.   DEATHS.

A	issued	944 and had 14; being equal to	15 per thousand.		
B	"	1901	"	27	"	14	"
C	"	838	"	11	"	13	"
D	"	2470	"	65	"	26	"
T P I	"	1596	"	12	"	$7\frac{1}{2}$	"

During the sixth year of its existence, only *two* deaths occurred out of the whole number of insurers in the Temperance Provident Institution, by which its annual average of mortality was reduced still lower.

70. Such comparisons, however, must not be regarded as demonstrating that the usual rate of mortality among "moderate" or "temperate" men, is reduced to half its amount by "total abstinence;" since other causes have doubtless concurred to keep down the mortality in the Temperance Provident Institution,—such as a more healthful condition of the class which has furnished most of the insurers — or a more favourable distribution of ages. But it

---

[1] *Endemic* — local; connected with particular localities or occupations.

will be seen to be impossible that either of these separately, or both conjointly, should have occasioned the whole of the difference above pointed out; the annual average, 6 per 1000, being no higher than that of the age of 15, which is more favourable than that of any other period of life. And we appear fully justified, therefore, in attributing a part of the result to the *abstinent* system practised by the Insurers in the Temperance Provident Office.

71. The influence of excess in the use of Alcoholic liquors in directly producing sickness and mortality, or in predisposing to it, is most remarkable in tropical climates, and especially in such as are otherwise unhealthy. It would seem, indeed, that the more unhealthy the station, the more freely do the residents at it indulge in the use of alcoholic stimulants; either from the mistaken idea that they enable them to withstand the effects of the climate, or from the desire that their life, if short, shall be a merry one. Some years since, the writer, being himself in the Island of St. Vincent in the West Indies, met with a gentleman resident in Tobago, who informed him that the average annual mortality amongst the Europeans of that island was about *one in three*. Upon inquiry into the habits of the residents, it was found that intemperance prevailed to a most fearful extent among them; few getting up in the morning without their glass of sangaree (wine and water), and the strength of their beverage gradually increasing during the day, until it arrived at neat brandy at night. He further spoke of it as no uncommon occurrence for a party of friends who had met at a drinking-bout, to be summoned within two or three days, to the funeral of one or two of their number. This gentleman was himself apparently quite indisposed to recognize between these occurrences any relation of cause and effect; being obviously under the belief that, if it were not for the protecting influence of good wine and brandy, his life would be worth a yet shorter purchase. We shall be led, however, by the evidence about to be adduced, to a different conclusion.

72. The writer has on various occasions sought for information from those who had preserved their health during a long residence in tropical climates, as to their habits in the use of Alcoholic liquors; and has almost invariably found that they had practised extreme moderation, if not total abstinence.

73. At the Statistical Section of the British Association, in the year 1848, a paper by Mr. Balfour having been read, on the "Means of maintaining the Health of Troops in India,"—in which paper the author attempted to show that Intemperance would be found to add but a small proportion to the deaths from climatorial diseases;[1] but that the special liability of Europeans to the diseases of hot

---

[1] *Climatorial diseases* — diseases dependent on climate.

climates arises from their unsuitability of constitution to any climate widely different from that of their own country,—an important discussion took place, in the course of which some valuable facts were established by the testimony of several officers present, (medical and otherwise) who had served in India and elsewhere. The returns contained in the paper, showed a marked difference in rate of mortality, between the ordinary Soldiers and the Officers; a difference which was greater according to the unhealthiness of the station Now a certain part of this difference must be admitted to be due to the superior character of the Officer's lodging, and to his partial exemption from the fatigue and the exposure to which the Soldier is liable. But the difference is chiefly to be accounted for by the difference in the manner of living between the Soldiers and the Officers; the former being allowed a regular ration of spirits, and many of them getting as much more as they can; whilst the latter are now comparatively abstemious, drinking wine or beer in place of spirits, and this to a much less extent than formerly. With regard to the Indian service, it was specially asserted by Lieutenant-Colonel Sykes, who has paid great attention to the Statistics of the Indian Army, that since it has become the custom among the Officers to drink bitter ale in place of wines or spirits, the rate of mortality among them is so greatly diminished, that promotion is no longer expected to take place more rapidly in the Indian Army than in other departments of the service. In illustration of the extreme injury done by Intemperance to the Indian troops, a Medical Officer stated that within a month after the arrival of the order for the discontinuance of Temperance Societies,[1] he had *forty* cases of Delirium Tremens in his own Regiment.

74. That the liability to climatorial disease is by no means inevitable, and that it is especially to be avoided by the adoption of the habits in regard to diet, &c., of the Native Population, where that is healthy,—is the testimony of all those who have had most extensive opportunities of forming a judgment on the subject. The two following citations from different publications,—the one by Lieutenant-Colonel Sykes, who was himself long resident in India,—the other by Dr. Daniell, Assistant Surgeon to the Forces, who has had the superintendence for a considerable time of some of the most unhealthy stations on the Western Coast of Africa,—will carry with them great weight. "I never followed a farinaceous or vegetable regimen myself in India," says Colonel S., "nor do I recommend it to others; but I ate moderately and drank little, and I have a

---

[1] The Authorities at the Horse Guards, who have taken the extraordinary step of putting down Temperance Societies in the Army, on the ground that every organization but the regimental is contrary to the discipline of the service, can scarcely be supposed cognizant of what they have to answer for.

strong conviction that much of European disease in India is trace-able to over-stimulus, and that the mortality among the European troops will not be lessened until the European Soldier is improved in his habits; until he is made to understand that temperance is for the benefit of his body, libraries for the benefit of his mind, exercise for the benefit of his health, and Savings' Banks for the benefit of his purse. *The climate of India is less to blame than individuals; for, in case Foreigners find the people in a country healthy, they should, to a certain extent, conform to the habits of the Natives to be healthy also.*[1] So with regard to Africa, Dr. Daniell says, " It is a well-known fact that the notorious insalubrity of Africa has frequently served as the scape-goat on which the blame of those evil consequences (resulting from the reprehensible indulgence of dissipated courses) might be unreservedly thrown, without the risk of their being disputed or questioned."[2] And again, when describing the Bight of Benin, one of the most pestilential localities on the surface of the globe, he says, " And yet, amid these regions so rife with disease and death, I have known Europeans reside for a number of years in the enjoyment of good health, from the simple secret of moderately conforming to the habits of the natives as regards their diet, exercise, and attention to the due performance of the cutaneous functions[3]."[4]

75. The evidence of Statistics, however, is more valuable on this point, than the mere affirmation of individuals, however trustworthy; and to this, as set before us by Colonel Sykes, we shall now proceed. — The per-centage annual mortality from sickness of the three armies of Bengal, Bombay, and Madras, for the last twenty years, has averaged as follows :—

	BENGAL.	BOMBAY.	MADRAS.
Native	1·79	1·291	2·095
European	7·38	5·071	3·846

This Table presents some very remarkable features. In the first place, the striking contrast between the rate of mortality of the European and of the Native Troops, serving together, and exposed to the same morbific causes. Secondly, the great difference between the mortality of the Troops serving in the different Presidencies. And thirdly, the circumstance that in the Madras Presidency the

---

[1] Vital Statistics of the Indian Army, in Journal of the Statistical Society, vol. x. p. 184.
[2] Sketches of the Medical Topography and Native Diseases of the Gulf of Guinea, Western Africa, p. 13.
[3] *Cutaneous functions* — functions of the skin.
[4] Op. cit. p. 61.

rate of mortality is *highest* amongst the Native Troops, and *lowest* among the Europeans.

76. Now on the first point Colonel Sykes remarks :—" I will not say that the question is absolutely solved by the reply ' Habits of Life ;' but I will say, reasoning from analogy, that the reply goes a great way to solve it. The European soldier in India is over-stimulated by food, over-stimulated by drink, and under-stimulated in mind and body. The European soldier eats a quantity of animal food every day of his life ; he drinks a quantity of alcohol every day of his life to the amount of a bottle of spirits in every five days, two drams being served out to him daily ; and he has not any mental and little bodily exercise. Happily the pernicious practice has been recently discontinued ; but time was, when the European soldier was compelled to take his dram by eight o'clock in the morning, with the thermometer varying from 70° to 90° or more, at the different seasons of the year, leaving him in a state of nervous irritation and thirst, which could only be relieved, as he thought, by further potations ; indeed I have been assured within the last few days by a pensioned Artillery Staff-Serjeant, *who never drank in India, and was only in hospital five days during twenty-one years' service*, that he has known, out of a detachment of 100 Artillery men, no less than eight men in straight-jackets at one time, absolutely mad from drink. Now animal food with the assistance of such an auxiliary, and combined with mental vacuity, go far to account for the excess of mortality amongst the Europeans."

77. The question next arises, why the mortality of the European troops in the Madras Presidency should be so much less than that of the others, being about *three-fourths* that of the Bombay troops, and but *little more than half* that of the Bengal army ; whilst, on the other hand, the mortality of the Native troops in the Bombay army is but *little more than two-thirds* of that of the Bengal army, and *less than two-thirds* that of the Madras army. There does not seem to be any such difference in the climatorial diseases, or in the character of the military stations of the three Presidencies, as are by any means sufficient to account for this discrepancy ; and if there were, we should expect them to manifest themselves alike in the Native and in the European army. That the reverse is the case, must be admitted to be a cogent argument, if not a complete proof, in favour of the insufficiency of any such account of the discrepancy. The following are the causes assigned by Colonel Sykes :—The *Bengal* European army has no supply of porter, but is furnished with rum, a spirit not so wholesome as arrack. On the other hand, the *Madras* army consume large quantities of porter, and drink comparatively little spirit, what they do consume being arrack. The *Bombay* troops have only recently commenced the consumption of porter, and the spirit they drink is understood to be more wholesome than rum, and less so

than arrack. "These results," says Colonel Sykes, "are certainly not conclusive; but I cannot help associating the increased consumption of malt liquor by the Madras Europeans with their comparative healthiness; and the gradations of the mortality in the Bengal and Bombay European troops as partly influenced by the quality (no doubt much more by the quantity) of the spirits they respectively consume."

78. On the other hand, the excess of mortality in the Native army of Madras above that of the Bengal and Bombay troops, is equally attributable to a difference in the habits of the individuals composing it. "Of the *Bombay* army," says Colonel Sykes, "six-eighths consist of Hindoos, and considerably more than half of the whole army are Hindostanees. These men never taste meat, fish, or spirituous liquors, but live, I may from personal observation venture to say, almost exclusively upon unleavened cakes of wheat or other cerealia, baked upon an iron dish, and eaten as soon as cooked. The great majority of the *Bengal* army consists of a similar class of men. The *Madras* army in its constituents is the reverse of the other two. In the cavalry there are from 6 to 7 Mussulmans to 1 Hindoo, and in the infantry there is 1 Mussulmau to every 1½ or 1¾ Hindoos; but amongst the latter there is a considerable number of low castes, without prejudices about food, and unrestrained by the prejudices of caste; therefore the majority of the Native troops of the Madras army can eat and drink like Europeans." Thus then we see, that whereas in the Madras army, in which the European and Native habits most closely assimilate, the mortality of the former is *less than double* (about 38 to 21) that of the latter; the mortality of the Bengal Europeans is *nearly six times* (about 74 to 13) that of the Bombay Natives; this difference bearing such a relation to the greater abstemiousness of the Native soldiers, and the larger consumption of spirits by the Europeans, that it is scarcely possible to avoid the inference that they must be connected in the relation of effect and cause.

79. The following returns are of value, as showing the proportion of sickness between the members of *Temperance Societies*[1] in the European Regiments serving in India, and the soldiers not members of these Societies; the average daily number of men in hospital of each class being stated, for each of the first six months of 1838, and the per-centage being calculated with reference to the strength of each division.

---

[1] In these Societies the pledge simply held the members to abstinence from distilled spirits, and from excess in the use of any fermented liquor.

7*

Months.	Strength of the Temperance Societies.	Strength of remainder of Regiment.	Relative proportions admitted to Strength.		Average daily per centage of men in Hospital.	
			Temperance Society.	Remainder of Regiment	Temperance Society.	Remainder of Regiment.
January	1953	2569	1 in 18·77	1 in 9·22	2·54	8·15
February	1840	2639	1 in 20·10	1 in 9·24	2·27	8·27
March .	1542	2879	1 in 14·44	1 in 7·14	2·94	8·66
April . .	1359	3081	1 in 10·9	1 in 5·26	5·47	10·28
May . .	1282	3161	1 in 18·44	1 in 6·35	5·24	10·66
June . .	1364	3075	1 in 19·53	1 in 6·37	4·55	10·35
Total	9340	17,354	1 in 16·47	1 in 7·28	3·65	10·20

Thus it appears that on the whole, the daily per centage of invalids among the members of the Temperance Societies was but 3·65, whilst in the remainder of the troops it was 10·20, or nearly three times as great.—The Cameronian Regiment, stationed in Fort William in the years 1837 and 1838, lost only *twenty-six* men in the first of these years, and *twenty-two* in the second; whereas the average mortality in Fort William, for a period of fourteen years previously, had been nearly *seventy-two*. The proportion of the Regiment which belonged to the Temperance Society was continually varying; but the general result of its operation was, that since the arrival of the Regiment in India, its annual consumption of spirits diminished from 10,000, 12,000, and even 14,000 gallons, to 2,516; the amount drunk in 1837 having been 9673 gallons less, and in 1838, 8242 gallons less, than the Regiment was entitled to draw. On the other hand, a considerable amount of beer and wine had been drunk; but these beverages are far less injurious to health, especially in India, than distilled spirits.

80. Having learned that the 84th Regiment of Her Majesty's Foot has for some time enjoyed the reputation of being one of the most temperate and well-conducted Regiments in the European portion of the Indian army, the writer has consulted the Army Medical Returns, for the purpose of ascertaining whether its rate of mortality has differed in any marked degree from the average given above; more especially since it has been quartered at Secunderabad, which lies under the bad repute of being one of the most unhealthy stations in the Madras Presidency. That this bad repute is well deserved, is shown by the fact that the annual mortality, for an average of fifteen years previously to 1846–7, has been 7·5 per cent.; *nearly double* the average of the whole Presidency, and *more than double* the average of the other stations. The evil seems traceable to the insufficient barrack-accommodation, rather than to the condition of the surrounding country; for one·

third of the men are obliged to sleep in the verandahs, and the re
mainder have by no means a due allowance of fresh air ; whilst
the officers of the regiment quartered there, and the Artillery company, who are better lodged, have not shown any excess of mortality
above the average.

81. Now in the year 1846–7, the average strength of H. M.
troops in the Madras Presidency was 5963, and the number of
deaths was 251, or 4·21 per cent. ; which is rather above the average mortality in this Presidency, calculated by Colonel Sykes from
the returns of twenty years. During the first eight months of this
period, the 84th Regiment was quartered at Fort St. George,
Madras, which is considered a healthy station ; it then performed a
march of between four and five hundred miles to Secunderabad, in
an unusually wet season, the roads (such as they were) being in
some parts knee-deep in water ;[1] and it took up its quarters at Secunderabad, about two months previously to the date of the return
(April 1st, 1847). The medical return of the regiment for this
year presents us with the almost unprecedently low number of *thir-
teen* deaths in an average strength of 1072 men ; the mortality
being thus at the rate of only 1·21 per cent.—Now during the
same period, the 63rd Regiment, which was quartered at Secunderabad up to February 1st, 1847, (or nine months out of the twelve)
lost *seventy-three* men, which was at the rate of 7·88 per cent. for
the entire year ; whilst the mortality for all *the other* stations in the
Madras command was only 3·02 per cent. for the same year.
Hence we see that the mortality of the 84th Regiment for the
year 1846–7, was only *two-fifths* of that of the average of the
healthier Stations in the Madras Presidency, which average its own
very low rate contributed to reduce.

82. During the year 1847–8, the total mortality in the Madras
Presidency was 227 to 6040 of average strength, or 3 76 per cent. ;
but this reduction from the preceding year was not due to any considerable difference in the rate of mortality at the other stations,
being almost entirely consequent upon the diminution in the number
of deaths at Secunderabad. For the 84th Regiment, which remained at that station during the whole year, lost in that time no
more than *thirty-nine* men out of an average strength of 1139, so
that its per-centage mortality was only 3·42 ; which was *below the
general average* of the Presidency, and less than half the average
rate at Secunderabad for fifteen years previously.—It seems impossible to attribute these remarkable results to anything but the abstinent habits of the soldiers of this regiment ; a large proportion
of them being *total abstainers,* and those who were not so being
very moderate in their consumption of alcoholic liquors.

---

[1] Of this march a more particular account will be given further on
(§ 140).

83. The comparison of the returns of this Regiment with those of others less remarkable for sobriety, affords a full confirmation of the deductions drawn from the statistics of crime in this country, as well as from individual observation, in regard to the influence of habitual Intemperance upon the moral conduct. That a large proportion of offences amenable to punishment, both in the civil population, and in the military and naval services, are committed under the direct excitement of Alcoholic liquors, there can be no kind of doubt; and the comparison of the insubordination and criminality of a drinking regiment, with the orderly and reputable conduct of an abstinent one circumstanced in other respects almost precisely the same, adds to the confidence with which we may assert that *Intemperance is the chief cause of Crime.* For besides the immediate provocation which alcoholic excitement may induce, it is indubitable that habitual excess has a tendency to debase the moral tone, and to weaken the controlling power of the will; — an effect for which the statements already made as to its action on the mind, through its organ the brain, leave us at no loss to account.[1]

---

[1] The author is happy to be able to add the testimony of Colonel Reid, who was for some years Governor of the Bermudas, and subsequently of the Windward Islands, in favour of the beneficial effects of Total Abstinence, in improving the physical condition, and in promoting the general welfare, of a tropical population composed of a mixture of Europeans with coloured people. To Colonel Reid belongs the high credit of being one of the few individuals occupying situations of high official responsibility, who have employed their influence in promoting this great measure of social reform. The author is informed by him, that the habit of Total Abstinence now prevails in the Bermudas to such an extent, that in some parishes there are no public-houses; the feeling of the people being so much against these, that no one will come forward to give the collateral security which is required from those who seek licenses for them.—In the Annual Reports of the Governors of Colonies which are laid before Parliament, and published, Colonel Reid speaks as follows, with reference to Barbadoes:—" I endeavoured, on my first arrival here, to revive a Temperance Society which had been before unsuccessfully attempted. This Society has now taken root in the midst of Distilleries; and promises to effect a great social revolution in West Indian habits.".

CHAPTER II.

DOES PHYSIOLOGY OR EXPERIENCE TEACH US THAT ALCOHOLIC LIQUORS SHOULD FORM PART OF THE ORDINARY SUSTENANCE OF MAN, PARTICULARLY UNDER CIRCUMSTANCES OF EXPOSURE TO SEVERE LABOUR, OR TO EXTREMES OF TEMPERATURE? OR, ON THE OTHER HAND, IS THERE REASON FOR BELIEVING THAT SUCH USE OF THEM IS NOT SANCTIONED BY THE PRINCIPLES OF SCIENCE, OR BY THE RESULTS OF PRACTICAL OBSERVATION?

THE reply to this question will be best furnished, in the opinion of the writer, by considering *seriatim*[1] how far science and experience lead to the belief, that the use of Alcoholic Liquors is advantageous, as fitting the system for the better endurance,—1st, Of severe *bodily exertion;*—2nd, Of severe *mental exertion;*—3rd, Of extreme *cold;*—4th, Of extreme *heat;*—5th, Of *morbific agencies.*[2] From the results of these enquiries it will be found not difficult to draw deductions as to the propriety, or otherwise, of making Alcoholic Liquors form part of the sustenance of Man under ordinary circumstances.

I. — ENDURANCE OF BODILY EXERTION.

84. All bodily exertion is performed by the instrumentality of the *muscular* apparatus, which is called into play by the agency of the *nervous* system. It is requisite, therefore, that we should begin by enquiring into the conditions under which their powers are respectively put forth; and the following may be stated as fundamental positions, in which all the most eminent Physiologists are now agreed.

I. Both the Nervous and Muscular systems require, for the energetic development and due maintenance of their respective powers, that their tissues shall be adequately supplied with the *Materials of growth and regeneration;*[3] whereby they shall be able to repair the effects of the loss, which every exercise of their vital endowments involves; and also to develope new tissue to meet increasing demands upon their functional activity.

II. The *functional activity*, both of the Nervous and Muscular systems, involves the *disintegration*[4] of a certain amount of their component tissues, by the agency of *Oxygen*: the evolution of their

---

[1] *Seriatim* — in order.
[2] *Morbific agencies* — agencies productive of disease.
[3] *Regeneration* — reformation; reconstruction.
[4] *Disintegration* — breaking up; a reduction to very minute particles.

peculiar forces being apparently dependent upon the return of the living tissue to the condition of dead matter, and upon the union of this matter with the oxygen supplied by the blood; whereby new compounds are formed, the retention of which in the circulating current would be detrimental to the vivifying qualities of the blood, and the continual elimination of which, therefore, is especially provided for. — Both these systems consequently require, as the condition of their highest activity, that they shall receive an adequate supply of blood, charged with Oxygen, and purified from the con taminating matters which it has taken up in the course of its circulation through the system.

III. For the fullest evolution of physical power, it is requisite that the Muscular system should receive an adequate excitation from the Nervous; and the amount of muscular force put forth on any occasion depends, *ceteris paribus*,[1] on the degree of nervous power which is caused to operate on the muscles, — a strong Emotional excitement, for example, being sometimes effectual in accomplishing that which the will could not effect.

85. If the Nerves and Muscles be inadequately nourished, it is impossible that their normal power can be developed, except under the influence of stimulants, and then only for a short time. If, on the other hand, the blood be imperfectly charged with Oxygen, it cannot supply a sufficient amount of that element for the performance of those chemical changes, which are involved in every action of the muscular and nervous apparatus. And if, besides being deficient in Oxygen, the blood be charged with carbonic acid, biliary matter, urea,[2] or other products of the disintegration of the body, the functional power of the nervous and muscular systems must undergo a marked diminution, in consequence of the deleterious influence which such matters exert upon their tissues.

86. Now it may be accepted as an indubitable fact in Organic Chemistry, that there is not the slightest relation of composition between Alcohol and Muscular tissue; and all our present knowledge of the subject tends to prove, that the albuminous[3][4] matters of the blood, which constitute the *pabulum*[5] of that tissue, cannot be generated within the body of man, or of any other animal, but are derived immediately from the food. We cannot regard Alcoholic liquors, then, as contributing to the nutrition of Muscular tissue; except in so far as they may contain albuminous matters in addition

---

[1] *Ceteris paribus* — other things being equal.

[2] *Urea* — a substance forming an essential part of the urine.

[3] *Albuminous* — of the nature of albumen; albumen is the nutritive portion of the blood.

[4] This term is here used to designate what are commonly known as the *protein* compounds; late researches having tended to show the incorrectness of the basis on which that appellation was founded.

[5] *Pabulum* — food, aliment, nutriment.

to the Alcohol, which is especially the case with "malt-liquors." But these matters would have the same nutrient power, if they were taken in the form of solid food.

87. We cannot speak with the same confidence, in regard to the *impossibility* of any assistance being afforded by Alcohol to the nutrition of the Nervous system ; since Nervous matter is essentially composed of fatty substances, which, though peculiar as containing a large quantity of phosphorus, do not seem to contain nitrogen ; [1] and since Alcohol is regarded by the Chemist as approximating the oleaginous [2] class of substances in its chemical relations.—But there are two circumstances, which render it highly *improbable* that Alcohol can ever be converted into Nervous matter. In the first place, we have no other example of an organic compound being found applicable to the nutrition of the animal tissues, which is the product of incipient *decay* or decomposition ; yet this may be affirmed to be the case with Alcohol, since the Alcoholic fermentation is the first of a series of degrading changes, which, if allowed to continue unchecked, terminates in the putrefactive process ; and we can scarcely imagine, therefore, that it can be an appropriate material for the formation of the most active and important part of the whole animal mechanism. Again, we have no other example of the application of an organic compound to the nutrition of the animal tissues, which exerts upon any of them such a decidedly *poisonous* influence in large doses, as we have seen to be exerted by Alcohol (§§ 13–16). The materials which constitute the *pabula* for the several tissues, are perfectly innocuous whilst they retain their normal constitution ; and their presence in the blood, in larger amount than usual, though it may in various modes be a source of functional derangement, never exercises any special deleterious influence upon the vital properties of the nervous, muscular, or any other tissue. On these grounds, then, it may be almost positively affirmed, that notwithstanding the chemical relation which Alcohol bears to Nervous matter, it cannot serve, either in its original condition, or under any other guise, as a pabulum for the generation of nervous tissue.

87. We seem justified by the laws of Physiology, therefore, in assuming that Alcoholic liquors cannot supply the first of the requi-

---

[1] It is usually stated, on the authority of Fremy, that the fatty acids of the nervous substance contain nitrogen ; this, however, is probably an error ; arising from the substance of the brain or nerves being submitted to analysis *en masse ;* for this substance consists not merely of the fatty contents of the cells and tubes, but of their albuminous walls ; and thus regarded chemically, it is a mixture of oleaginous with a small quantity of albuminous matter, which last, when included in the analysis, would give to the former ingredient the appearance of containing azote. — (See Valentin's Lehrbuch der Physiologie, Band. I., p. 174.)

[2] *Oleaginous* — oily or fatty.

sites already enumerated for the development of the physical power of the nervous and muscular apparatus; and we have next to consider what is its capacity in regard to the second. It may be safely affirmed that the introduction of Alcohol into the blood cannot stand in the place of the Oxygen which is essential to the functional activity of the nervous and muscular systems; on the contrary, its presence in the blood would rather tend to impede the oxidation of their organic components, both by the more cogent demand for oxygen which it will itself set up, and also by the preventive influence which it is well known to exercise over the oxidation of other organic substances (§§ 117, 118). In both these modes, it will not only interfere with that action of the Oxygen of the blood upon the Nervous and Muscular substances, which is essential to their functional activity; but it will also tend to check the removal by oxygenation, of those products of decomposition, whose continuance in the blood is attended with most serious injury to the system. In so far, in fact, as the presence of alcohol in the circulating current tends to give to arterial blood a venous character, it must thereby impair its power of serving as the exciting fluid (for so we may term it) of the nervous and muscular battery. And this it does in the first instance, by obstructing the elimination of Carbonic Acid, as will be shown hereafter (§ 118); but more remotely, by that interference with the proper functional activity of the Liver and Kidneys, which we have seen to be among the most ordinary consequences of the free and habitual use of Alcoholic liquors (§§ 54–58).

88. But although we are led by the preceding considerations to regard the *regular* employment of Alcoholic liquors as rather a detriment than an aid to the development of nervo-muscular power, there is a third point towards which we have to direct our inquiry; namely, whether the peculiar *stimulating* effect of Alcohol, which is especially exerted upon the Nervous system, may not enable a greater amount of nervous energy to be produced, and a greater amount of muscular power to be thereby called forth, than could be generated without its aid. In considering this question, it is most important to keep in view the difference between a *temporary* and a *sustained* effort. We have seen that the usual effect of a moderate dose of Alcohol is, in the first instance, an increase in the force and rapidity of the Circulation, and in the activity and energy of the functions of the Nervous system; and both these conditions will be favourable to the development of Muscular power, so long as they continue. But such a state cannot long endure. We may increase the amount of Nervous power developed in a given time, by the influence of Alcoholic stimulants; or we may prolong its generation by the same kind of assistance, when it would otherwise have failed. But as every exertion of Nervous power, like that of Muscular, involves the death and decay of a certain amount of the tissue by

which it is evolved, there is a limit to the possibility of its generation; so that we find the continuance, or even the increase, of the
stimulus ceasing after a while to produce any effect; and the exhausted power can only be recovered by a lengthened period of
repose, which shall allow time and opportunity for the regenerating
processes to be performed, at the expense of nutrient material drawn
from the blood. Until this has been effectually accomplished, the
Nervous power is at least as much *below* par, as it previously was
above it; so that the loss is certainly equivalent to the gain. And
the more the Nervous system has been forced, by the influence of
Alcoholic stimulants, to give forth its powers beyond their natural
limit, whether as to duration or intensity, the greater will be the
degree and duration of that subsequent depression, which speaks so
unmistakeably of the need of rest and reparation.

89. Hence, therefore, we should anticipate that although the use
of Alcoholic stimulants may enable a greater amount of physical
force to be put forth within a given time, than could otherwise be
generated, they can be of no assistance in the *sustentation* of nervo-
muscular power; and if the previous considerations be also taken
into the account, we should be led to expect that, in the long run,
severe bodily labour will be better borne without alcoholic stimulants than with them,—provided always that the digestive apparatus
be in good working order, and be adequate to prepare that amount
of alimentary material, which is required for the regeneration of the
tissues disintegrated by use.

90. We have now to inquire how far the results of practical experience are coincident with these theoretical views; and whether
it is found on actual trial, that complete abstinence from Alcoholic
liquors is favourable, or the 'reverse, to the endurance of severe
bodily labour. It cannot be denied that the ideas current, among
the labouring classes more especially, as to the teachings of experience on this point, are opposed to our theoretical deductions. But
there are many circumstances which should lead us to mistrust the
popular voice on such a question, and to seek for proofs of a kind
that may be more firmly relied on. The "universal experience"
of former generations might be quoted in favour of a multitude of
absurd notions, which we now treat as simply ridiculous; and when
there is this additional complication, that the liking for alcoholic
liquors is such as very readily to make "the wish father to the
thought," we find an additional ground for suspicion. But the
chief cause of our mistrust is this,—that there is no appreciation in
the popular mind, of the connexion between the *immediate* and the
remote effects of Alcoholic stimulants. A glass of malt liquor, or
a small quantity of spirits, repeated three or four times a day, is
found to increase the bodily vigour for a time; and this increase is
set down as so much positive gain, no account being taken of the

subsequent depression, which is considered as ordinary fatigue. Evidence of this kind is therefore of little or no value; and the only facts that can be admitted as having any weight, are those which bring into comparison the total amount of labour executed *with* and *without* the aid of fermented liquors, during lengthened periods of severe toil; these being the indications, not of the amount of force which may be temporarily set forth, but of that which can be habitually exerted; and therefore of the general vigour of the system, rather than of its power in a state of excitement.

It would be easy to cite several *modern* testimonies to the superiority of the Abstinence principle (to say nothing of the ancient ones), from Benjamin Franklin down to Dr. Forbes ; — the former of whom tells us in his Autobiography, that he was accustomed, when working as a Pressman in a London printing-house, and taking only bread and water instead of the porter which his companions drank (as they said) to acquire strength for their work, to carry a large form of letters in *each* hand up and down stairs, to the astonishment of his porter-drinking companions, who found *one* of them a sufficient load ;—whilst the latter assures us that in a recent excursion amidst the mountains and valleys of Switzerland, which were chiefly traversed on foot, he found his own " sexagenarian" [1] vigour, sustained by cold water only, quite on a par with that of his younger companions, who indulged in a moderate allowance of wine. [2] Such examples, however, might be regarded as exceptional, because individual; and as affording no contradiction of the supposed general result of experience. They prove, however, that there is nothing positively incompatible, in the habit of total abstinence from alcoholic liquors, with the sustenance of a high degree of nervo-muscular power. And it may be well to fortify this position with a few additional testimonies, relating to cases in which the power of endurance was very severely tried.

92. Thus, a nail-maker at Glasgow assured the writer, that after five years' experience of the abstinence system, he "found hard work easier, and long hours more readily to be endured;" and that being a member of the Fire Brigade, he was on one occasion called upon for continuous exertion for *seventy-three* hours, which he endured, with no other beverage than coffee and ginger-beer, while all his comrades were "beat and fell away." The following statement, forwarded to the writer from Leeds, was signed by thirty-four men engaged in laborious employments; out of whom twelve belonged to the class whose occupations are commonly regarded as peculiarly trying, seven of them being furnace-men at foundries and gas-works, two of them sawyers, one a whitesmith, one a glass-blower,

---

[1] *Sexagenarian* — one sixty years old.
[2] " Physician's Holiday," p. 26 and passim.

and the last a railway guard "We, the undersigned, having practised the principles of total abstinence from all intoxicating liquors during periods ranging from one to ten years, and having, during that time, been engaged in very laborious occupations, voluntarily testify that we are able to perform our toil with greater ease and satisfaction to ourselves (and we believe more to the satisfaction of our employers also) than when we drank *moderately* of these liquors; our general health and circumstances have also been considerably improved." [1] With regard to harvest-work, again, which is extremely trying to the strength, both from the continuity of the exertion required, and the heat of the weather at the time of its performance, there is ample testimony that those who go through it upon the abstinence principle, are better able to sustain it, than those who endeavour to support their strength upon fermented liquors; and that if an adequate supply of nutritious food be provided for them, the former will even *increase* in weight, whilst going through this severe toil. In some parts of the county of Cornwall, where the 'abstinence' system is more extensively practised than in any other county in England, it is the general practice to get in the harvests without any allowance of fermented liquors; many labourers, who are habitually moderate drinkers, feeling the benefit of the 'abstinence system' at such times.

93. The following example, drawn from another source, is of peculiar value, as showing the comparative effect of the two systems upon the same individuals.—The writer was acquainted, some years since, with a gentleman who had been for some years at sea in the Merchant-service, and who not long previously had commanded a vessel during a voyage from New South Wales to England. After passing the Cape of Good Hope, the ship had sprung so bad a leak, as to require the continued labour, not merely of the crew, but also of the officers and passengers, to keep her afloat during the remainder of her voyage, a period of nearly three months. At first the men were greatly fatigued, at the termination of their "spell" at the pumps; and after drinking their allowance of grog, would "turn in," without taking a proper supply of nourishment. The consequence was, that their vigour was decidedly diminishing, and their feeling of fatigue increasing, as might be expected on the principles already laid down. By the directions of their Commander (who although very moderate in his own habits, at the time of the writer's acquaintance with him, was by no means a disciple of the Total Abstinence school, which renders his testimony the more valuable), the allowance of grog was discontinued, and coffee and cocoa were substituted for it; a hot "mess" of these beverages being provided, with the biscuit and meat, at the conclusion of

---

[1] See Appendix B.

every watch. It was then found that the men felt inclined for a good meal of the latter, when the more direct but less effective refreshment of the alcoholic liquor was withdrawn; their vigour returned; their fatigue diminished; and after twelve weeks of incessant and severe labour (with no interval longer than four hours), the ship was brought into port with all on board of her in as good condition as they ever were in their lives.

94. Numerous examples might be cited of *comparative trials* between two sets of labourers, as nearly as possible alike in other respects, but the one practising Total Abstinence, whilst the other has relied upon the assistance of Alcoholic liquors. So far as the writer is aware, all these contests have given results in favour of the abstinence system, when the period of the experiment has been sufficiently protracted to give its merits a fair trial; and although it may be asserted that such results are one-sided, as having been made known to the public by the professed advocates of a *system*, yet considering the very large interests involved in the maintenance of the existing state of things in regard to the use of fermented liquors, it might be reasonably expected that their upholders would make known to the world any results of an opposite description, had they really occurred. The following statement, furnished to the writer by a gentleman at Uxbridge, has the advantage of being the comparative return of the *regular labour* of a whole year, performed by two sets of men, the one working on the "abstinent," the other on the "moderate" system, but not pitted against each other in a contest for victory. It relates to brick-making, which is commonly accounted one of the most laborious of all out-door employments. "Out of upwards of twenty-three millions of bricks made in 1841, by the largest maker in the neighbourhood, the average per man made by the beer-drinkers in the season was 760,-269; whilst the average for the teetotalers was 795,400—which is 35,131 in favour of the latter. The highest number made by a beer-drinker was 880,000; the highest number made by a teetotaler was 890,000; leaving 10,000 in favour of the teetotaler. The lowest number made by a beer-drinker was 659,500; the lowest number made by a teetotaler was 746,000; leaving 87,000 in favour of the teetotaler. Satisfactory as the account appears, I believe it would have been much more so, if the teetotalers could have obtained the whole 'gang' of abstainers; as they were very frequently hindered by the drinking of some of the gang; and when the order is thus broken, the work cannot go on."

95. The experience of *large bodies* of men, which becomes matter of public notoriety, is in many respects preferable, as demonstrating (to say the least) the perfect compatibility of Abstinence from alcoholic liquors with the highest degree of physical vigour, and with the greatest power of endurance of bodily labour. Thus, almost

every traveller who has visited Constantinople, has been struck with the remarkable muscular powers of the men engaged in the laborious out-door employments of that city.   Mr. W. Fairbairn, an eminent machine-maker at Manchester, remarked that "the boatmen or rowers to the caiques, who are perhaps the first rowers in the world, drink nothing but water; and they drink profusely during the hot months of summer.   The boatmen and water-carriers of Constantinople are decidedly, in my opinion, the finest men in Europe as regards their physical development, and they are all water-drinkers." [1]   And several other observers bear testimony to the extraordinary strength of the porters of Constantinople, who are accustomed to carry loads far heavier than English porters would undertake, even under the stimulus of alcoholic beverages; yet these Turkish porters never drink anything stronger than coffee.

96.   The following statement, made upon the authority of Mr. Tremenhere, one of the Commissioners employed to report on the state of the Mining population, shows how completely the doctrines of the reputedly "universal experience," in regard to the support afforded by Alcoholic liquors to the laborious artisan, are negatived by the results of a change of habit, forced upon those most unwilling to adopt it.   "A remarkable and most satisfactory instance," says Mr. T., "of a successful attempt to put a check upon the indulgence in ardent spirits, has occurred at the iron-works of Messrs. Houldsworth of Coltness, employing about eight hundred colliers, miners, furnace-men, &c.   Much loss and annoyance had frequently been occasioned by the negligent or wilful misconduct of workmen under the influence of this habit; and the Messrs. Houldsworth, having in vain endeavoured to put an end to it by persuasion and advice, resolved to do what they could by removing the temptation. They accordingly, about three years ago, forbade the sale of spirits at the store, and at the inn at their works, and ordered that the furnace-men should not be allowed to drink spirits during their hours of labour.   These men had been accustomed to drink four or five glasses of whiskey during each 'shift,' in addition to what they might choose to drink at their own homes.   They remonstrated strongly, and affirmed that it was impossible for them to do their work without this quantity of whiskey.   They were not long, however, before they found their error; they now drink nothing but water during their work, and tea and coffee at their meals; what they spent on whiskey they now spend in wholesome and nutritious food; they allow that they do their work better, and that the change has been a great blessing to themselves and their families; and that it is the best thing that ever happened to them.   I was after-

_____

[1] Sanitary Report, 1840, p. 252.

8 *

wards informed that among the colliers and miners, there wiↄ a marked improvement from the same cause."

97. The experiment has now been carried on upon a still larger scale, for many years, amongst the seamen of the Merchant-service, both of this country and the United States; and the result has been, with few exceptions, so favourable to the Abstinence principle, that it is now adopted by a very large proportion of American trading vessels; to whose general superiority in equipment and management over the mercantile marine of this country, in the greater part of which the employment of alcoholic liquors is still continued, a large body of evidence was recently given before a Committee of the House of Commons. The exceptions just alluded to relate to the reputed liability of "temperance" seamen to suffer from endemic[1] or malarious[2] diseases. Into this point we shall enquire under a subsequent head; and the writer thinks that it will there appear that this liability, if it really exist, results from the deficiency of any measures that shall serve as a substitute for the alcoholic stimulus, in rendering the system less obnoxious to the influence of the poisonous emanations to which it is exposed. (§§ 145–147.)

98. The writer has had the opportunity of ascertaining from Shipowners who have adopted the "temperance" system (which on board ship, is equivalent to "total abstinence"—no other alcoholic liquor being substituted for spirits), that they have found no difficulty in obtaining the services of excellent seamen, when a fair compensation is made in the superior quality of the provisions and allowances, or in the rate of wages, for the "stopping of the grog." In fact, such ships are in positive request among seamen of the best character; proving that in spite of the well-known attachment of their class to spirituous liquors, they are sensible of the advantages of habitual abstinence from them. The writer having himself made a voyage to the West Indies and back, some years since, in a "temperance ship," had the opportunity of remarking that during a heavy gale of nearly three days' duration, which was continually taxing to the uttermost the strength of a crew far too small for the size of the ship, the men were at least as ready for the renewal of their exertions, as they would have been if supported by alcoholic stimulants; whilst in various rowing-matches, which took place between them and the crews of other ships, whilst lying in port, they were generally the victors. This last fact is not cited as proving the superiority of the abstinence system, since the difference might be attributed to the superior physical power of the crew; but it sufficiently indicates that there is in the "abstinence system" *nothing unfavour-*

---

[1] *Endemic* — resulting from local causes.
[2] *Malarious* — resulting from a bad state of the air.

able to the maintenance of that power during the vicissitudes of a seaman's life.

99. The following extract from a private letter from a Captain in the Merchant-service, contained in the "*Adviser*" for October, 1849, gives a valuable testimony in favour of this conclusion:—"I harboured in Newfoundland on the 23d of December last, the coldest day that had been registered there for the last six years, the thermometer on shore indicating twenty degrees below Zero. I can honestly say, it was the most severe frost I ever was in on the water, during the twenty-nine years that I have been employed in the Newfoundland trade. I remained on shore from the time mentioned above, until the 2d of March, and then embarked for Brazil, where, in April, we had the thermometer ranging from 80° to 87°, and remained in that climate till the middle of July All that time, the whole of my crew, with two exceptions, were strictly teetotal, and all able to eat their allowance, and do their share of hard work, in the sun and out of it, taking in and out cargo. The two exceptions did, in one solitary instance, infringe the law; and they paid the penalty in severe headache and debility for some days."

100. It is rare for any occasion to present itself, during the march of an army, of testing the power of sustaining this kind of prolonged exertion, without the supposed assistance derived from the use of Alcoholic liquors; but opportunities of this kind have occurred, the abstinence having been in some instances voluntary, whilst in others it was compulsory; and the results have in both cases been most completely confirmatory of the principles formerly laid down. Two of the most striking examples of this kind within the writer's knowledge, will be mentioned hereafter under the head of "Endurance of Heat" (§§ 140, 141); these marches having been performed under an elevation of temperature which rendered them peculiarly trying. And he will here confine himself to the mention of the fact, that during Sir John Moore's retreat to Corunna, the army was found to *improve* in health and vigour, as soon as the usual allowance of spirits was unattainable. This fact is the more remarkable, as the circumstances under which this march was performed must have been peculiarly depressing to the feelings of the men, and could not but have operated unfavourably (according to the invariable experience of retreating armies) upon their physical powers.

101. The experience of whole nations, previously to the introduction of Alcoholic liquors amongst them, is equally in favour of the assertion, that prolonged and severe muscular exertion may be at least as well borne without their assistance as with it. Where, for example, shall we meet with greater power of endurance of toil, than was displayed by the North American Indians in "following a trail," before their race became deteriorated by European vices?

102. The question, it may again be remarked in conclusion, is not to be decided by the amount of strength which may be put forth at a single effort.   It may be freely admitted that when the body is exhausted by fatigue, an Alcoholic stimulus, which excites the Nervous system to increased exertion, may impart a temporary strength, which shall enable the next effort to be successful in doing that which could not have been accomplished without it.   But there is reason to believe that the power of *sustained* exertion is thereby impaired; and that those who have habitual recourse to this stimulus are really doing themselves more harm than good. This will be most assuredly the case, when they allow it to take the place of the solid food, which their nervous and muscular systems require for their regeneration; and the tendency of the habitual employment of alcoholic liquors, when the body has been fatigued with severe and prolonged muscular exertion, is generally to diminish rather than to increase the desire for solid aliment,—as the examples above quoted clearly indicate.   And as it is the latter alone that can afford real and permanent support, it is obvious that any habit which diminishes the natural disposition to profit by it, *must* be positively injurious in its influence upon the bodily vigour.

II. — ENDURANCE OF MENTAL EXERTION.

103. All that has been said of the influence of Alcoholic liquors on the development and sustentation of Physical force, will equally apply to Mental power; since, whatever may be our views as to the relation between Mind and Matter, it is not now questioned by any Physiologist, that the Brain is the *instrument* by which all mental power is exercised, in Man's present state of existence, and that the continued development of this power' is consequently dependent upon those conditions, which are favourable to the maintenance of the functional activity of the Nervous system in general. These conditions we have seen to be (I)—The healthy nutrition of the nervous substance; and (II)—The due supply of highly oxygenated and depurated blood.'   The former process is certainly not *dependent* upon the assistance of alcoholic liquors; and from the considerations already urged (§ 87), it seems in the highest degree improbable that they can be of the least advantage to it.   The latter cannot in any degree be improved, but must be rather impaired, by the use of fermented liquors; which, as already stated, tends to deteriorate the quality of the blood, and to obstruct its oxygenation.

104. That the use of Alcoholic stimulants, however, is attended in most persons with a *temporary* excitation of mental activity,—

' *Depurated blood* — blood from which all refuse, deleterious and worn-out matters have been removed lighting up the scintillations of genius into a brilliant flame, or assisting in the prolongation of mental effort when the powers of the nervous system would otherwise be exhausted,—may be freely conceded; and it is upon such evidence as this, that the common idea is based, that it *supports* the system under the endurance of mental labour. This idea, however, is probably as erroneous, as the no less prevalent fallacy of regarding alcoholic liquors as capable of increasing the power of physical exertion. No physiological fact is better established, than that of the depression of the mental energy consequent upon the undue excitement of it, by whatever causes that excitement may have been occasioned; and the rapid and brilliant flow of thought which may have been called forth by the alcoholic stimulants, gives place, usually after a few hours, to the opposite state of languor and despondency.

105. The influence of Alcoholic stimulants seems to be chiefly exerted in exciting the activity of the *creating* and *combining* powers; such as gives rise to poetical imaginations, to artistic conceptions, or to the sallies of wit or humour. It is not to be wondered at, then, that men possessing such powers, should have recourse to alcoholic stimulants as a means of procuring a temporary exaltation of them; and of escaping from the fits of depression to which most persons are subject, in whom the imaginative and emotional tendencies are predominant. Nor is it to be denied, that many of those mental productions which are most strongly marked by the inspiration of genius, have been thrown off under the stimulating influence of alcoholic liquors. But, on the other hand, it cannot be doubted that the depression consequent upon the high degree of mental excitement which is thus produced, is peculiarly great in such individuals, completely destroying for a time the power of mental effort; and hence it does not at all follow, that either the authors of the productions in question, or the world at large, have really benefited thereby. Moreover, it is the testimony of general experience, that where men of genius have habitually had recourse to alcoholic stimulants for the excitement of their powers, they have died at an early age, as if in consequence of the premature exhaustion of their nervous energy; Mozart, Burns, and Byron may be cited as remarkable examples of this result. Hence, although their light may have burned with a brighter glow, like a combustible substance in an atmosphere of oxygen, the consumption of material is more rapid; and though it may have shone with a soberer lustre without such aid, we cannot but believe that it would have been steadier and less prematurely quenched.

106. We do not usually find that the men most distinguished for that combination of intellectual powers which is known as *talent*, are disposed to make such use of Alcoholic stimulants for the purpose of augmenting their mental powers; for that spontaneous ac

tivity of the mind itself, which it is the tendency of alcohol to excite, is not favourable to the exercise of the observing and purely reasoning faculties, or to the steady devotement of concentrated attention to any subject which it is desired to investigate profoundly. Of this we have a remarkable illustration in the habits of practised gamblers; who, when about to engage in contests requiring the keenest observation and the most sagacious calculations, and involving an important stake, always "keep themselves cool," either by entire abstinence from fermented liquors, or by the use of those of the weakest kind in very small quantities. And we find that the greatest part of that intellectual labour which has most extended the domain of human knowledge, has been performed by men of remarkable sobriety of habit, many of them having been constant water-drinkers. Under this last category, it is said,[1] may be ranked Demosthenes and Haller; Dr. Johnson in the latter part of his life took nothing stronger than tea, while Voltaire and Fontenelle used coffee; and Newton and Hobbes were accustomed to solace, not to excite, themselves with the fumes of tobacco. In regard to LOCKE, whose long life was devoted to constant intellectual labour, and who appears, independently of his eminence in his special objects of pursuit, to have been one of the best-informed men of his time, the following very explicit and remarkable testimony is borne by one who knew him well. "His diet was the same as other people's, except that he usually drank nothing but water; and he thought that his abstinence in this respect had preserved his life so long, although his constitution was so weak."[2]

107. Having for several years past been himself performing an amount of steady mental labour, which to most persons would appear excessive, the writer may be allowed to refer to his own experience, which is altogether in favour of Total Abstinence from alcoholic liquors, as a means of sustaining the power of performing it. Having been brought up as a water-drinker, he never accustomed himself to the habitual use of alcoholic liquors; scarcely ever tasting them, except when occasionally led to do so by social influences, or when he believed that a small amount of stimulus would improve the "tone" of his system, which is liable to a peculiar relaxation in certain states of the atmosphere. On determining, about four years since, to give up the occasional use of wine, &c., as a social indulgence, he still held himself free to employ it when he might think it likely to increase the general powers of his system; and for some time he continued to have occasional recourse to alcoholic stimulants (never exceeding a single glass of wine, or half a tumbler of bitter ale), when he felt him-

---

[1] Macnish's Anatomy of Drunkenness, p. 36.
[2] Life by Lord King, vol. ii. p. 60.

self suffering under the peculiar depression just referred to. He gradually, however, found reason to doubt the utility of the remedy; and has for the last two years entirely given it up. During these two years, he has gone through a larger amount of mental labour than he ever did before in the same period of time; and he does not hesitate to say, that he has performed it with more ease to himself than on his former system; and that he has been more free than ever from those states of depression of mental energy, which he was accustomed to regard as indicating the need of a temporary support to antagonize the depressing cause. In fact, he now finds that when these do occur, the use of alcoholic stimulants (taken even in very small amount) is decidedly injurious to him; diminishing, rather than increasing, his power of mental exertion at the time; and leaving him still less disposed for it after their effect has gone off. He attributes this change to his entire disuse of alcoholic liquors under all other circumstances; and he cannot but believe that the results which he now experiences, and which have led him to relinquish these stimulants altogether, are the *natural* effect of them upon the healthy system; and that the benefit which some persons consider themselves as deriving from their use, arises from their simply removing *for a time* the depression which results (at a long interval it may be) from their *previous* employment.

108. Two remarkable cases have recently fallen within the author's knowledge, in which individuals leading a life of considerable intellectual exertion, and long habituated to the moderate use of stimulants, have derived considerable benefit from their relinquishment.—In one of these cases, a pint of ale was the usual daily allowance; to which a little spirits and water at night was occasionally added. The relinquishment was commenced as an experiment, and without any intention of persevering should it not succeed; but the benefit has been so great, that the abstinence has been subsequently continued as a settled practice. This gentleman not only finds his general health improved, but declares that his power of intellectual exertion is *much greater* than formerly; and in particular, that he finds himself quite fresh and ready for work in the morning, instead of losing time, as formerly, in bringing himself up to the point at which he left off, the night before. — In the other instance, the usual daily allowance was from two to four glasses of wine; and this was affirmed to be necessary to *keep down* a state of mental excitement to which the individual was subject, and to brace the mind to steady exertion. Failing health, however, having occasioned a recourse for a time to the Hydropathic treatment, this gentleman, on returning in a state of renewed vigour to his usual avocations, wisely determined to persevere in the disuse

of stimulants; and he has since continued to practise the abstinent system, with great benefit to his bodily and mental health.

109. Even if we admit, however, that a certain amount of mental labour may be performed with more facility in a limited time, under the moderate use of Alcoholic stimulants, it is still questionable, whether we do not, on the whole, rather lose than gain by their employment. For if they cannot afford *pabulum*[1] for the formation of nervous matter, and if their influence is exerted rather in producing its disintegration than its growth, its *destruction* rather than its *construction*, it follows that every excess of exertion performed under their influence must be followed by a correspondingly long period of incapacity, during which the regenerating processes have to be performed, and the brain again fitted for the discharge of its duties; and if it should be forced into activity before this renovation has been duly performed, the amount of stimulus required to bring it up to the working point will be greater, and all the consequent evils increased. These theoretical predictions are, it is believed, in full accordance with what observation teaches with respect to the results of reliance upon alcoholic stimulants for support during mental labour; although, for obvious reasons, it is not possible to obtain the same pointed and decisive evidence on this topic, as in regard to the endurance of physical exertion, or of extremes of temperature. — But it is frequently urged, that although the use of Alcoholic liquors to produce a stimulating action upon the *brain*, is injurious, yet that benefit is derived from the employment of a quantity sufficient to stimulate the *stomach* to the proper discharge of its duties, by digesting that amount of food which the system requires, but which the exhaustion of nervous power prevents it from duly appropriating. This doctrine, which equally applies to the state of imperfect indigestion resulting from other causes, will be better considered when we have inquired into the reputed efficiency of Alcoholic liquors in supporting the system under exposure to the extremes of Cold and Heat, to which question we have next to proceed.

III. — ENDURANCE OF COLD.

110. The power of Alcoholic liquors to enable the body to resist the depressing influence of external Cold, is, perhaps, the best established of all its attributes, not merely in the estimation of the uninformed public, but in the opinion of those who have scientifically considered the question. This is by no means surprising. The genial warmth which is experienced for a time, when a glass of spirits is taken on a cold day, appears to afford unmistakeable

---

[1] *Pabulum* -- nutriment; material adapted to form or sustain.

evidence of its heat-producing power; and the chemical properties of Alcohol would seem to indicate, that, under such circumstances, it does not merely act as a stimulant, increasing the activity of the circulation, and augmenting the nervous energy — but that it also affords the material for that combustive process by which the heat of the body is sustained in a form peculiarly suitable for rapid and energetic appropriation to this purpose.   The authority of Liebig is continually quoted in support of this view; but more has been built upon his statements than they legitimately support.   For his arguments are rather directed to prove that Alcohol cannot become a pabulum for the tissues, and that its sole use, therefore, must be in maintaining the temperature of the body by the combustive process, than to show that it is superior to other materials, to whose employment, as they exert no stimulating influence, the objection raised against alcohol cannot apply.   That we may place this question upon its proper basis, it will be necessary to consider the circumstances under which the combustive process is usually carried on.

111. That the maintenance of Animal Heat is chiefly, at least, dependent upon the union of the carbon and hydrogen of certain materials contained in the blood, with oxygen taken in by the lungs; — and that the non-azotized [1] ingredients of the food are specially appropriated to this purpose; — are positions in which there is now such a general agreement amongst Physiologists, that they may be assumed as a basis for our further inquiries.   The non-azotized ingredients of ordinary food may be grouped under two heads — the *saccharine*,[2] and the *oleaginous*;[3] the former including all those farinaceous matters which can be converted into sugar; and the latter consisting of oil and fat in every form.   The former may be considered as *hydrates of carbon*; their proportionals of oxygen and hydrogen being such as to form water; so that in combustion they will only consume as much oxygen, as will convert their carbon into carbonic acid.   On the other hand, the proportion of oxygen in the latter is comparatively small; so that in combustion they require as much, as will not only convert their carbon into carbonic acid, but will also unite with that part of the hydrogen for which no equivalents of oxygen previously exist in the compound. Thus an equivalent of Starch consists of 12 Carbon, 9 Hydrogen, and 9 Oxygen; whilst an equivalent of Stearine, the basis of the solid fats, consists of 136 Carbon, 132 Hydrogen, and 10 Oxygen

---

[1] *Non-azotized* — destitute of nitrogen.
[2] *Saccharine* — resembling sugar.
[3] *Oleaginous* — oily; fatty.

Multiplying the numbers of the former by 11¾, so as to bring them better into comparison with the latter,[1] we find that,—

11¾ equiv. of Starch =: 140 Carbon, 105 Hydrogen, 105 Oxygen.
1 equiv. of Stearine = 136 Carbon, 132 Hydrogen, 10 Oxygen.

Now in the former case, the number of equivalents of oxygen necessary for the conversion of the starch into carbonic acid and water, will be no more than that required for the change of its carbon into carbonic acid, namely, $(140 \times 2 =)$ 280; but in the latter case, the number required will not be merely that which will convert the 136 eq. of carbon into carbonic acid, namely, $(136 \times 2 =)$ 272; but also that required for combination with those 122 equivalents of hydrogen, for which no equivalent of oxygen exists in the compound, making in all $(272 + 122 =)$ 394. A much more energetic combustive process is required, therefore, for the conversion of Stearine into carbonic acid and water, than for effecting the same conversion upon Starch; for not only is the quantity of free oxygen consumed much larger, but the amount of heat generated will be much greater; since much more heat is produced by the combustion of Hydrogen, than by that of Carbon.

112. Now the atomic composition of Alcohol being 4 equivalents Carbon, 6 eq. Hydrogen, and 2 eq. Oxygen, it is intermediate, in regard to its proportion of oxygen, between the farinaceous and the oleaginous substances; bearing, however, a strong resemblance to the latter, in regard to the large proportion of Hydrogen for which it does not contain an equivalent of oxygen. This will be best seen by multiplying the equivalent of Alcohol by 31½, which will bring the total weight of its carbon and hydrogen exactly to the same figure with that of 11¾ of Starch.[2]

31½ equiv. of Alcohol = 126 Carbon, 189 Hydrogen, 63 Oxygen.

Thus, then, in the combustion of this amount of Alcohol, there will

---

[1] We thus make the sum total of the weights of Carbon and Hydrogen very nearly the same in the two cases; for

$$140 \text{ equiv. of Carbon} (140 \times 6) = 840$$
$$\text{and } 105 \text{ equiv. of Hydrogen } (105 \times 1) = 105$$

$$945$$

$$\text{whilst } 136 \text{ equiv. of Carbon} (136 \times 6) = 816$$
$$\text{and } 132 \text{ equiv. of Hydrogen } (132 \times 1) = 132$$

$$948$$

[2] For 126 equivalents of Carbon .... (126×6) · 756
and 189 equivalents of Hydrogen .... = 189

$$945$$

not merely be required ($126 \times 2 =$) 252 equiv. of oxygen, for the conversion of its carbon into carbonic acid; but as ($189 - 63$) 126 equivalents of hydrogen exist in the compound without any equivalent of oxygen, that number of equivalents of oxygen will be required to convert all the hydrogen into water, making in all ($252 + 126 =$) 378. This amount is not far from that required by an equivalent quantity of Stearine; and as a much larger proportion of it is consumed by the hydrogen, it is obvious that the heat produced must be greater than that evolved by the combustion of an equal weight of hydro-carbon contained in the latter.

113. Considered, therefore, merely in the light of *fuel*, Alcohol is superior to Oleaginous substances, whilst it is of far higher value than any substance of the Saccharine group; and of this the Chemist is practically aware, for he finds that a spirit-lamp gives more heat than an oil-lamp. Were the human body simply a lamp or a furnace, therefore, we should have no room for doubt as to the efficiency of Alcohol in maintaining its heat; and it is because the influence of Alcohol upon the *vital* functions is too much disregarded, its share in the mere *chemical* process of combustion being too exclusively kept in view, that notions are entertained of its value, which are contradicted by lengthened and extended experience. This will be best understood, if we examine in the first instance into the circumstances under which other non-azotized substances taken in as food are made to contribute to the maintenance of Heat.

114. Of such substances, a certain amount is usually circulating in the blood. All analyses indicate the existence of *Fatty* matters in that fluid; their proportion, however, varies considerably, being much greater after a full meal of which oleaginous matters have formed a part. Although the amount is usually not too large to be held in solution by the alkali of the blood, yet the serum of blood drawn within a few hours after such a meal, is usually found to be rendered opaque or milky, by the presence of an unusual quantity of oleaginous particles suspended in it, in a state of very fine division; these however gradually diminish in amount; and in a few hours more, the serum becomes clear again, indicating that these particles have been in some way disposed of. This, we can scarcely doubt, is the consequence of their having been " burned off" by the respiratory process, which is every hour carrying away at least one-third of an ounce of carbon from the blood.—The evidence that the *Saccharine* elements of the food are used-up in the respiratory process with equal rapidity, is not quite so distinct; since these elements appear to be usually introduced into the blood in the condition of lactic acid,[1] the detection of which is attended with some uncertainty. But it has been sufficiently proved that when the saccha-

---

[1] *Lactic acid* — an acid obtained from milk.

rine ingredients of the food are unusually abundant, they enter *as
such* into the blood; where they may be detected shortly after a
meal, especially if that meal have been preceded by a long fast.
Like the superfluous fatty matters, however, they soon disappear;
being carried off, there can be little doubt, by the respiratory process.

115. In this manner, then, the Heat-producing materials are
usually supplied to the system from meal to meal; the greater part
of them being destined for elimination from the blood within a short
time after their admission into it; and the power of sustaining heat
existing in its greatest vigour, only whilst some of them remain
unconsumed.   This inference is confirmed by ordinary experience;
for every one knows how much more severely Cold is felt after a
fast of some hours' duration, than after a full meal.   We are accustomed to refer the difference to the condition of the stomach; but
the stomach may have been emptied, by the completion of the
digestive process, long before the increased susceptibility to cold
commences; so that it would be more correct to refer this increase
to the exhaustion of the supply of combustible material last introduced into the blood, than to the vacuity of the stomach.   That an increase in the power of maintaining heat should be almost immediately
produced after the ingestion of food into the stomach, is to be accounted for, not merely by that augmented activity and energy of
the general circulation which accompanies the digestive process, but
also by the rapidity with which nutrient matters find their way into
the blood,—the turbidity of the serum, consequent upon the introduction of fatty substances, having been observed as early as half-
an-hour after the meal of which they have formed part.[1]

116. The admission of these matters into the current of the circulation cannot be discovered to produce any effect upon the system
in general, otherwise than by sustaining the temperature of the
body.   In fact, they seem to be the legitimate *pabulum* for the *combustive* process, just as Albuminous matters constitute the *pabulum*
for the *formative* processes whereby the tissues are generated.
When they are present in excess, the superfluity is withdrawn by
the production of adipose tissue; which stores up the fatty matters
for future use.   When, on the other hand, the supply is not equivalent to the consumption required for the maintenance of the heat of
the body, the fatty matters which are among the normal constituents
of the blood are first drawn upon; and as the proportion of these
is diminished, it is supplied from the contents of the cells of adipose[2] tissue.   In this manner, the animal temperature is kept up

---

[1] See, for the experiments on which several of the foregoing statements
are founded, the paper of Drs. Buchanan and R. L. Thomson, in the Medical Gazette, Oct. 10th, 1845.

[2] *Adipose* — fatty.

nearly to its usual standard, even in spite of the total deprivation of food, so long as unconsumed fatty matter remains in the body; but death then speedily takes place, in consequence of the cooling of the body, unless the temperature be sustained by external warmth. And death may result also from the subjection of the body to a very low temperature, whilst there is still much fatty matter left in the tissues; as if this matter could not be re-introduced into the circulating current, with sufficient rapidity to supply the demand for an extraordinary quantity of heat-producing pabulum. Further, when the store of fatty matter has been entirely exhausted, and the animal has nothing whatever to fall back upon, it is requisite that the supplies of new material introduced into the system should suffer no intermission; for immediately that they are exhausted, the temperature of the body begins to fall, and death speedily supervenes unless a fresh supply be afforded.[1]

117. We are now prepared, then, to inquire into the question, how far Alcohol may be advantageously employed habitually as a heat-producing material; and whether there are any peculiar or extraordinary circumstances, under which it is to be preferred to others. And as one means of arriving at the truth on this point, we must examine more particularly into the influence of the introduction of alcohol into the blood, upon the respiratory process. For our knowledge upon this point, we are chiefly indebted to the experiments of Dr. Prout and to Vierordt. The former states[2] that Alcohol and all liquors containing it which he had tried, have the remarkable power of *diminishing* the quantity of carbonic acid gas in the expired air, much more than any thing else which he had made the subject of experiment; this effect being most decided, when the liquor was taken upon an empty stomach. The latter[3] fully confirms Dr. Prout's observations; having found that, in four experiments, the per-centage of carbonic acid fell, after from half to a whole bottle of wine had been taken, from 4·54 to 4·01; and that this effect lasted between one and two hours. He further found that, when he drank wine with his dinner, the usual increase in the per-centage of carbonic acid expired after a full meal did not take place.

118. These facts are of great importance. For although it may be very possible that, as suggested by Liebig, the increased formation of *water*, which will occur when Alcohol is the combustive material, compensates for the diminution in the amount of carbonic acid expired, and thus the normal amount of heat may be generated, —yet there are clear indications that, when thus present in the

---

[1] For the experiments on which the foregoing statements are founded, see the work of M. Chossat, entitled "Recherches sur l'Inanition.

[2] Annals of Philosophy, vols. ii. and iv.

[3] Physiologie des Athmens, &c.

blood, with other materials which ought to be excreted, alcohol exerts an injurious influence, by retarding their combustion. This it will do in two ways; first, by taking their place as the more readily combustible material; and secondly, in virtue of the anti-septic [1] influence which it exerts upon other substances, preventing or retarding chemical changes in them. That such is the case, appears from the experiments of Bouchardat; who found that, when alcohol is introduced into the system in excess, the blood in the arteries presents the aspect of venous blood, showing that it has been prevented from undergoing the proper oxygenating [2] process.[3] And the experiments of Dr. Prout afford additional support to this conclusion; for he observed that no sooner had the effects of the alcohol passed off [4] (which they did in his case with frequent yawn-ings, and a sensation as if he had just awoke from sleep), than the amount of carbonic acid exhaled *rises much above* the natural standard — thus giving, it would seem, unequivocal evidence of the previous abnormal retention of carbonaceous matter in the system.

119. From the foregoing considerations, then, we may conclude that the effects of Alcohol as a heat-producing material, will only be advantageously experienced, when the blood does not contain a supply of other matters waiting for removal by the respiratory pro-cess; and this, we believe, will be found entirely conformable to experience; the greatest assistance being derived from it, when the body is exposed for a time to severe cold, after a long previous fast, and when, for the reasons already given, the heat-producing power is much less than usual, even although there should be no lack of material stored up in the body. This is well illustrated by the following incident, which Dr. Macnish relates of himself.[5] "I was travelling on the top of the Caledonian Coach, during an intensely cold day, towards the end of November, 1821. We left Inverness at five in the morning, when it was nearly pitch dark, and when the thermometer probably stood at 18° Fahr. I was disappointed of an inside seat, and was obliged to take one on the top, where there were nine outside passengers besides myself, mostly sportsmen re-turning from their campaigns in the moors. From being obliged to get up so early, and without having taken any refreshment, the cold was truly dreadful, and set fear-noughts, fur-caps, and hosiery, alike

---

[1] *Antiseptic* — preventing the occurrence of putrefaction.

[2] *Oxygenating* — supplying with oxygen, which is the vitalizing principle of the blood.

[3] This result has been also noticed as a consequence of the inhalation of the vapours of Ether and of Chloroform, which are allied to Alcohol in com-position and properties; and in cases in which the state of Anæsthesia has been very profound, the temperature of the body has undergone a con-siderable depression.

[4] Sailors can generally tell when the " grog is out of them."

[5] Anatomy of Drunkenness, p. 307.

at defiance. So situated, and whirling along at the rate of nearly nine miles an hour, with a keen east wind blowing upon us from the snow-covered hills, I do not exaggerate when I say that some of us, at least, owed our lives to ardent spirits. The cold was so unsufferable, that on arriving at the first stage we were nearly frozen to death. Our feet were perfectly benumbed; and our hands, fortified as they were with warm gloves, little better. Under such circumstances, we all instinctively called for spirits, and took a glass each of raw whiskey, and a little bread. The effect was perfectly magical; the heat diffused itself over the system, and we continued comparatively warm and comfortable, till our arrival at Aviemore Inn, where we breakfasted. This practice was repeated several times during the journey, and always with the same good effect. When at any time the cold became excessive, we had recourse to our dram, which insured us warmth and comfort for the next twelve or fourteen miles, without on any occasion, producing the slightest feeling of intoxication. Nor had the spirits which we took, any bad effects either upon the other passengers or myself. On the contrary, we were all, so far as I could learn, much the better for it; nor can there be a doubt, that without spirits, or some other stimulating liquor, the consequences of such severe weather would have been highly prejudicial to most of us." — This last statement cannot be admitted without an important reservation, sufficient to invalidate any inference drawn from this or similar cases as to the *necessity* for alcoholic liquors for the maintenance of the animal heat under exposure to severe cold. For it will be observed that the party started on their journey after a fast of several hours, no food having been taken that morning; and there is every reason to believe that if Dr. Macnish and his companions had breakfasted heartily before the commencement of their journey, they would not have found it necessary to have had such frequent recourse to the spirit-bottle; easily-digested solid food, especially such as includes oleaginous matter, taken in conjunction with hot liquids (especially Coffee), being at least as efficacious as a heat-producing material, as alcoholic liquors can be. In proof of this assertion, we shall now cite a series of facts which are, we conceive, quite adequate to demonstrate it.

120. In the first place, the author may relate his own experience of a journey performed on the outside of a stage-coach from Exeter to Bristol, on the 20th of January, 1838, a day memorable for the severity of its temperature, and for that remarkable prediction of the occurrence which gave a temporary celebrity to "Murphy's Almanack." The traveller, as in the preceding case, was "whirled along at the rate of nearly nine miles an hour" (which in these days of railroad speed must be accounted but a snail's pace); and though not exposed to "a keen east wind from the snow-covered

hills," was subjected to a much lower atmospheric temperature, the thermometer having stood during the day at 8°, or twenty-four degrees below freezing point.   Having fortified himself with a hearty breakfast, however, and having been in some measure previously inured to the cold by a severe frost of a fortnight's duration, he did not suffer from it to any extraordinary degree ; and with the aid of a fresh supply of food at dinner, he arrived at his journey's end without any greater degree of numbness of the extremities, than a short exposure to the genial warmth of a good fire subsequently removed.   No fermented liquor was taken by the writer on this journey; and he cannot think that he could have derived any other benefit from it, than that, by accelerating the general circulation, it might have possibly kept up a more rapid flow of blood through the surface and extremities.   But this would have been a doubtful benefit, if at the same time the combustion of the materials supplied by the food had been retarded by the presence of the alcohol in the blood.

121. The writer has heard many of the now almost extinct race of Stage-coachmen,—who had been induced to give up their former habit of imbibing a glass of ale or brandy-and-water at every stage, and to substitute an occasional cup of hot coffee and a rasher of . toasted bacon,—speak so decidedly in favour of the superior efficacy of the latter system, that he doubts if any man who had the resolution to adopt it, ever returned to his old habits except from the love of liquor.

121. Experience on a much larger scale, and under a greater severity of cold, leads to the same conclusions.   The Esquimaux, Greenlander, or Canadian, relies upon his solid aliment, which contains a considerable amount of oleaginous matter, for his power of resisting cold ; and when amply supplied with food, does not dread the exposure of his person to cold of the greatest severity.   Thus Captain Parry mentions with surprise that he saw an Esquimaux female uncover her bosom, and give her child suck, in the open air, when its temperature was *forty degrees below zero*.   And Sir J. Richardson, in a letter to the writer, states that "plenty of food and good digestion are the best sources of heat," and that "a Canadian with seven or eight pounds of good beef or venison in his stomach, will resist the greatest degree of natural cold, in the open air, and thinly clad, if there be not a strong wind."   The inhabitants of the Arctic regions appear to have a natural relish for the very oleaginous food, which Nature has provided for them, in the whales, seals, bears, and other animals upon which they chiefly subsist; and this taste is acquired by Europeans when exposed to the same conditions.   Thus Dr. King, who accompanied Sir George Back in his over-land expedition in search of Sir John Ross, informed the author that whereas he had been previously accustomed to reject every particle of fat, owing to the dislike he felt for it, he

found himself able, during his Arctic journey, to eat any amount of it with relish, and even experienced a positive craving for it; and his experience led him to consider himself as far better fortified against the cold by the use of an oleaginous diet, than by that of fermented liquors.—Testimony to the same effect is given by Dr. J. D. Hooker, who was one of the medical officers in the Antarctic expedition under the command of Sir James Ross.   He says, in a letter to the author,—"Several of the men on board our ship, and amongst them some of the best, never touched grog during one or more of the antarctic cruises.   They were not one whit the worse for their abstinence, but enjoyed the same perfect health that all the crew did throughout the four years' voyage. Many of our men laid in large stocks of Coffee; and when practicable had it made for them after the watch on deck. These men, I believe, would willingly have given up their spirits in exchange for coffee; but we could not ensure them the latter on the requisite occasions."

122. The foregoing statements appear sufficient to prove that a sufficient supply of Heat-producing food effects all that can be attributed to Alcoholic liquors in sustaining the heat of the body; but we shall now go further, and endeavour to establish the position, that the use of alcoholic liquors is positively injurious, when the exposure to cold is prolonged, and especially when muscular exertion is required. Thus Dr. Hooker says, in the communication just cited;—"I do think that the use of spirits in cold weather is generally prejudicial. I speak from my own experience. It is very pleasant. The glass of grog warms the mouth, the throat, and the abdomen; and this, when one is wet and cold, with no fire, and just before turning into damp blankets, is very enticing. But it never did me one atom of good; the extremities are not warmed by it; and when a continuance of exertion or endurance is called for, the spirit does harm, *for then you are colder or more fatigued a quarter or half an hour after it, than you would have been without it.*" The testimony of others who have been subjected to still more trying exposure, is to the same effect. Thus Sir J. Richardson states, as the result of his most severe experience,—"I am quite satisfied that spirituous liquors, though they give a temporary stimulus, diminish the power of resisting cold. We found on our Northern journey, that tea was much more refreshing than wine or spirits, which we soon ceased to care for, while the craving for the tea increased. Liebig, I believe, considers that spirits are necessary to northern nations, to diminish the waste of the solids of the body, and that tea is less useful; but my experience leads me to a contrary conclusion." Dr. King's testimony was precisely to the same effect. In fact it would appear that very general concurrence exists on this point among all those qualified to form an unprejudiced judgment in regard to it; since we find that in all the

recent over-land Arctic expeditions, sent out by the British Government, it has been expressly provided that no fermented liquors shall be used by the parties who proceed upon them; and that the Hudson's Bay Company have for many years entirely excluded spirits from the fur countries to the north, over which they have exclusive control, "to the great improvement," as Sir J. Richardson states, "of the health and morals of their Canadian servants and of the Indian tribes."[1]

123. That such are the teachings of sufficiently prolonged experience, not merely in the frigid zone, but wherever the same conditions present themselves, will appear from the two following statements.—It is mentioned by Dr. Forbes,[2] as the result of his personal enquiries from the guides at Chamouni, that when they are out upon their winter expeditions among the Alpine snows, they never find it advantageous to take any thing stronger than the weak wines of the country ; considering the use of spirits to be decidedly inimical to their power of sustaining exertion in an atmosphere of very low temperature.—The writer had the opportunity, about a twelvemonth since, of conversing with a very intelligent man of above seventy years of age, residing at Wareham in Dorsetshire, who had spent more than fifty winters as a fowler; in which vocation he had been exposed to the utmost severity of the winter's cold ; since it can of course be most profitably pursued, when the largest number of birds are driven southwards by the intensity of the frost in their northern residence. He stated that he had frequently been out for a fortnight at a time, without lying down save in his little boat, and scarcely ever obtaining warmth from a fire during that period ; and notwithstanding such severe trials, he was a remarkably hale and vigorous man for his years. Being himself the proprietor of a small public-house, he cannot be supposed to have any prejudice against the use of fermented liquors, in which he indulges in moderation ; but his testimony to the writer was most explicit to the following effect ;—

---

[1] To the above testimony, the author may add the following, with which Mr. Eaton has favoured him.—The Rev. Richard Knill, for many years a Missionary at Petersburgh, stated in a Public Meeting, in regard to the delusion which prompted people to use ardent spirits "to keep out the cold," that the Russians had long since found out the injurious effects of taking them in very cold weather. When a regiment was about to march, orders were issued over-night that no spirits were to be taken on the following morning; and to ascertain as far as possible that the order had been complied with, it was the practice of their officials answering to our corporals, carefully to smell the breath of every man when assembled in the morning before marching, and those who were found to have taken spirits were forthwith ordered out of the ranks, and prevented from marching on that day ; it having been found that such men were peculiarly subject to be "frost bitten," and otherwise injured.

[2] Physician's Holiday, p. 26. note

that although the use of ale or brandy might seem beneficial in causing the cold to be less felt at first, (so that when out for no more than a day or two, he did not think it necessary to abstain from it,) the case was quite reversed when the duration of the exposure was prolonged; the cold being then most severely felt, the larger was the proportion of fermented liquors taken. And he further stated, that all the fowlers of his acquaintance, who had been accustomed to employ brandy with any freedom, whilst out on prolonged expeditions, had died early; he and his brother (who had practised the same abstinence as himself) having outlived nearly all their con temporaries.

124. Hence it may be argued upon scientific principles, that whilst the use of Alcoholic liquors may for a time afford assistance in maintaining the heat of the body, so as the better to enable it to resist the influence of severe cold, they have no such advantage over Oleaginous matter, in affording a *pabulum* for the respiratory process, as sufficiently compensates for their injurious effect in preventing or retarding the oxygenation of those ingredients of venous blood, which ought to be continually eliminated by the respiratory process. Consequently, looking at the Chemical influence of alcohol merely, we might expect the prolonged employment of alcoholic liquors to induce such a vitiation of the blood, as will impair its fitness for the manifold purposes which it is destined to answer. No such result will follow the ingestion of heat-sustaining food; since this waits its time for the combustive operation, without interfering with the oxygenation of other matters; and if not itself consumed, it is stored up within the body until the time of need. But again, although the stimulating effect of alcoholic liquors is less during the exposure to cold than it is under ordinary circumstances, yet it cannot be altogether prevented by the more rapid combustion which the alcohol undergoes; and it might be anticipated, therefore, from what we know of the general action of stimulants, that the depression which follows upon their use would render the body peculiarly obnoxious to the influence of cold; so that, although they may help to keep up the temperature of the body *for a time*, by imparting increased energy to the circulation, yet when that energy is succeeded (as it must be sooner or later) by the opposite condition, the cold will be felt with greater intensity.

125. The predictions thus based on Physiological principles, are found, as we have seen, to be in most perfect harmony with Experience. For this teaches in the first place, that although Alcoholic liquors may afford advantages equal or even superior, regarded simply as material for the combustive process, to those derivable from solid food, those advantages are not of long duration; so that, for enabling the body to resist the continued influence of severe cold, alcoholic liquors are far inferior in potency to solid food

And, secondly, that although the increase in the energy of the circulation, resulting from the stimulating effect of alcoholic liquors, may prevent the depressing influence of the cold from having its ordinary action upon the system, provided that it be exerted only whilst that effect lasts; yet that after it has subsided, the cold is felt with augmented severity, and its action upon the system is proportionately injurious.[1]

126. The question whether there are *any* circumstances under which the use of Alcoholic liquors can be positively advantageous for the purpose of enabling the body to resist Cold, will be considered in the succeeding chapter (§§ 182–187).

IV. — ENDURANCE OF HEAT.

126. Having thus concluded our enquiry how far the use of Alcoholic liquors is necessary or desirable for arming the body against the depressing effects of cold, we shall consider their agency in supporting the system under the enervating influence of extreme Heat. The belief in the existence of such an agency is scarcely less strongly or generally entertained, than that of their protective power against cold; but it must be manifestly due, if it exist, to

---

[1] The author has preferred basing his conclusions upon information which he has obtained by his personal enquiries. He might easily have brought together a considerable amount of published testimony to the same effect. The following statements, contained in the work, entitled "Bacchus," are in complete harmony with those which he has himself adduced. "In 1619, the crew of a Danish ship of sixty men, well supplied with provision and ardent spirit, attempted to pass the winter at Hudson's Bay; but fifty-eight of them died before the spring: while in the case of an English crew of twenty-two men, in the same circumstances, but destitute of distilled spirit, only two died. In another instance of eight Englishmen, also without spirituous liquors, who wintered in the same bay, the whole survived and returned to England; and four Russians left without ardent spirits or provisions, in Spitzbergen, lived for a period of six years, and were at length restored to their country. In the winter of 1796, a vessel was wrecked on an island off the coast of Massachusetts; there were seven persons on board; it was night; five of them resolved to quit the wreck and seek shelter on shore. To prepare for the attempt, four of them drank freely of spirits; the fifth would drink none. They all leaped into the water; one was drowned before he reached the shore; the other four came to land, and in a deep snow and piercing cold, directed their course to a distant light. All that drank spirits failed, and stopped, and froze, one after another; the man that drank none reached the house, and about two years ago was still alive." (p 374.) The evidence of Captain (now the Rev. Dr.) Scoresby, who was for many years the captain of a whaling ship, is precisely to the same effect with that of the Arctic travellers whose testimony has been already cited. He gives it as his decided opinion that spirits are injurious in cold climates; and speaks of the reaction as especially pernicious, in diminishing the power of sustaining cold, as well as that of muscular exertion.

&ome *modus operandi* different from that which renders them ser-
viceable in the opposite condition. For it cannot be imagined that
they can be of any service by affording pabulum for the combustive
process, when that process is already generating more heat than
the body, exposed to a high external temperature, can possibly
need. Nor can it be supposed that the loss of the *watery* portion
of the blood, by the perspiratory process, can be in any degree repaired by the ingestion of *alcoholic* liquids. It must be presumed,
then, that whatever energy their use may communicate to the body,
must be derived from their stimulating properties; and must be
subject to there disadvantages, which are inseparable from the
habitual employment of stimulants. Each of these points, however, requires a fuller examination.

127. It is well known to the Physiologist, that the Respiratory
process in warm-blooded animals is much less energetic at high
temperatures than at low; the system having in itself the power
of regulating the amount of matter which it shall burn off, in
order that its heat may be kept up to the proper standard. Thus
it was ascertained by the experiments of Letellier,[1] that the amount
of carbonic acid set free by Birds, when they are breathing in an
atmosphere of from 86° to 106° Fahr., is scarcely more than *one-
third* of that which they generate in an atmosphere of 32°; and
by similar experiments upon small Mammalia,[2] it was ascertained
that they only give off, between 86° and 106°, about *half* as much
carbonic acid, and between 59° and 68° about *two-thirds* as much,
as they generate at 32°. The experiments of Vierordt[3] upon his
own person lead to a similar conclusion in regard to Man; although
the difference is not so great. For he states that the average
amount of carbonic acid exhaled by him per minute, between the
temperatures of 24° and 47° Fahr., was 18¼ cubic inches; whilst
the average between the temperatures of 66° and 92° was but 15¾
cubic inches. It is obvious, then, that the demand for combustive
material at high temperatures must be comparatively small; and
that the residents in hot countries cannot require the same supply
of heat-producing aliment, as is needed by the inhabitants of the
frigid zone. We see this indicated in the quality of the non-
azotized[4] material which Nature has provided for their use; for
whilst the dwellers amid the Arctic and Antarctic seas derive their
chief sustenance from these *oleaginous* articles which have the
greatest heat-producing power, the vast population of the Equatorial

---

[1] Comptes Rendus Tom. xx. p. 795; and Ann. de Chim. et de Phys.
Tom. xiii. p. 478.
[2] *Mammalia* — animals that suckle their young.
[3] Op. cit. §§ 73–82.
[4] *Non-azotized* — destitute of nitrogen.

region derives its principal support from these farinaceous,[1] vege-table products, whose non-azotized portion, belonging to the *sac-charine* class, has the lowest calorific agency (§ 111).

128. It is very necessary, however, to bear in mind, that the Respiratory process is not one of simple calorification;[2] for it is one of the most important of all those excretory operations, whereby the *waste* or effete[3] matter of the system is eliminated[4] from the blood. This, in fact, may be regarded as the essential part of the function, which is common to all animals; the combustion of an additional amount of hydro-carbonaceous matter, for the purpose of maintaining the temperature of the body at a fixed standard, being peculiar to the warm-blooded classes. It is evident, then, that from the diminution of the total quantity of carbonic acid exhaled at high temperatures, the *excretory*[5] part of the respiratory function will be more liable, than at low or moderate temperatures, to inter-ference from any agency which still further checks the oxygenation of the combustible matter of the blood.

129. Now as we have found that, under exposure to severe Cold, the stimulating effects of Alcoholic liquors (especially when taken at intervals, in small quantities at a time), are but little felt, the alcohol being burned off before it can accumulate so as to exert any considerable influence on the Nervous system; — so might we ex-pect that, under the influence of external Heat, when the combustive process is greatly reduced in activity, the stimulant effects of alcohol should be more rapidly produced and more powerfully exerted. And further, if the views formerly stated be correct as to the effects of the absorption of alcohol into the blood, in preventing the elimi-nation of matters which ought to be carried off by the respiratory process, we should expect that the use. of alcoholic liquors in warm climates would exert this obstructive influence in a peculiar degree.— Both these anticipations are confirmed by ample experience, which thus bears testimony to the soundness of our principles. For it is well known that a far smaller quantity of alcoholic liquor suffices to produce intoxication beneath a burning sun than in a frosty atmo-sphere; so that individuals who are not aware of this fact sometimes become intoxicated, without having exceeded the allowance which they believed to be perfectly compatible with sobriety. Again, it has been continually observed that when Alcoholic liquors are taken during the performance of severe labour in an extremely high tem-

---

[1] *Farinaceous* — of the nature of meal or flour.
[2] *Calorification* — producing heat.
[3] *Effete* — refuse matter; worn-out matter.
[4] *Eliminated* — separated, removed.
[5] *Excretory* — adapted to the removal and expulsion from the body of all such matters as are no longer adapted for its support, or are of an injurious nature.

perature, their temporary stimulation is followed by a very rapid and decided failure both of nervous and muscular power; so that men who drink largely of such liquors in the intervals of their work, are obliged to abstain from them whilst their labour is in progress. This result appears fairly attributable to vitiation of the circulating blood, consequent upon the retention of matters destined for excretion; the removal of which, by the oxygenating process, has been obstructed by the presence of alcohol.   And the same inference appears legitimately deducible from the peculiar tendency (already referred to, §§ 54, 55,) which the habitual use of alcoholic liquors in warm climates has to engender diseases of the Liver; the duty of separating those hydro-carbonaceous products of the waste of the system, which are poisonous if retained in the blood, being unduly thrown upon the liver, when their elimination by the lungs is interfered with.

130. That the use of Alcohol is especially necessary to support the system under its excessive loss by Perspiration at high temperatures, is an idea so commonly held, that it demands a serious refutation; although the fallacy of the notion, that because *water* is drawn off from the blood through the pores of the skin, *alcohol* must be taken into the stomach to replace it, would appear self-evident.   The fundamental error seems to lie in the notion, that copious perspiration in itself really weakens the system; whilst it is, in fact, nothing else than the means by which the external warmth is prevented from raising the heat of the body above its normal standard.   The determination of the blood to the skin, which that heat excites, appears to cause an unusual transudation [1] of the watery part of the blood through the thin-walled capillaries [2] of the sweat-glands; just as certain diuretic [3] medicines increase the quantity of water in the urine, by causing an increased determination of blood to the kidneys; but with this large amount of watery fluid, very little solid matter passes off, — none, in fact, but what is purely excrementitious.

131. That Perspiration, however abundant, has in itself no weakening effect, — except by diminishing the quantity of water in the blood (which is readily supplied by absorption from the stomach),— appears from the fact, that if persons exposed to a very high temperature make no bodily exertion, they do not experience any loss of vigour, if copiously supplied with cold water.   In fact, such exposure may be made to conduce very decidedly to the invigoration of the system.   All travellers who have tried the Russian baths, speak of the feelings of renovation which the copious perspiration:

---

[1] *Transudation* — passing through as perspiration.
[2] *Capillaries* — very minute vessels, like hairs.
[3] *Diuretic* — that which increases the secretion of urine.

and the subsequent plunge into cold water, produce in the wearied frame.   And those who have given a fair trial to the Hydropathic treatment, in *appropriate cases*, are unanimous in the same testimony. The writer has himself been in a stove-room, in which delicate females were accustomed to remain for half-an-hour or more, when it was heated to a temperature of from 140° to 170° Fahr.; their wrappings becoming saturated by copious perspiration, the material for which was supplied by the water administered to them internally from time to time; and he has had ample assurance to the effect, that this process, when followed by the cold plunge, had usually an invigorating influence, which quite sets aside the idea that the act of perspiration is in itself exhausting, or that it removes from the system any thing which it can be requisite for alcohol to supply.

132. The peculiar fatigue which usually results from muscular exertion at a high temperature, is generally set down as the consequence of the excessive perspiration; although the fact is, that the fatigue is chiefly to be attributed to the interference with the vaporous or "insensible" transpiration, which is produced by the accumulation of liquid or "sensible" perspiration on the surface of the skin, and by the saturation of the garments in contact with it.   For the same fatigue is experienced when the atmosphere is loaded with dampness, even at a low temperature; and it has been the uniform result of the attempt to use any muscular effort, when the body has been clothed in water-proof garments made after the fashion of ordinary clothes, so as not only to keep out the rain, but to keep in the insensible perspiration.   In either case the effect is the same;— the due vaporization of fluid at the surface of the skin is checked; the cooling influence of the perspiration is not exerted; and the heat of the body itself is injuriously augmented.[1]   And as an augmentation of from 11° to 13° in the temperature of a warm-blooded animal produces an invariably *fatal* result, so can it be readily understood that an increase of 2° or 3° must be attended with injurious consequences, so long as it lasts.

133. Among these consequences we may probably rank a still further diminution in the quantity of carbonic acid exhaled from the lungs; as well as an obstruction to the cutaneous respiration,[2]

---

[1] Thus it was found by MM. Delaroche and Berger, that when animals were exposed to the temperature of 120°, their bodies being enveloped in close boxes, whilst their heads were free, a thermometer placed in the mouth showed an increase of 6° in the heat of the body, in the course of seventeen minutes; this elevation being obviously due to the obstruction to the transpiration from the surface of the body.   When by continued exposure to a heated atmosphere saturated with moisture, the temperature of the body was raised from 11° to 13° above the natural standard, the animals uniformly died.

[2] *Cutaneous respiration* — discharge of vaporous and gaseous matter by the skin.

which, although its proportional amount has not yet been satisfactorily ascertained, is certainly of no mean importance in the depuration[1] of the blood. Hence an accumulation of excrementitious[2] matters will take place in the circulating fluid, such as affords quite a sufficient explanation of the peculiar fatigue which is experienced when muscular exertion is called for in a heated atmosphere already charged with moisture. And we should expect that such exertion could be performed with much less feeling of exhaustion in an atmosphere of *dry* air, though of very high temperature,—such as that of glass-houses, gas-works, or foundries,—than in the less heated atmosphere of tropical countries, which usually contains a considerable amount of watery vapour. This is undoubtedly the fact; and as a far larger amount of liquid will be carried off by insensible transpiration[3] in the former case than in the latter, it proves the correctness of our position, that it is *not* the loss of liquid from the skin, which is the cause of the peculiar exhaustion that results from muscular exertion in a heated atmosphere;[4] and that we are to look for the source of that exhaustion in the elevation of the temperature of the body itself, which will be produced with peculiar facility in a damp and heated atmosphere; and in the accumulation of excrementitious matters in the blood, which will be especially likely to take place when their elimination[5] through the lungs is being checked, at the same time that an increased amount is being generated by the *waste* of the muscular tissues.

134. If, then, our fundamental positions have been just, and our argument correct, we should infer that, putting aside their peculiar influence upon the nervous system, the use of Alcoholic liquors during muscular exertion in a heated atmosphere, and especially when that atmosphere is charged with moisture, can be nothing else than injurious; as tending to interfere still more with that elimination of excrementitious matters from the blood, which is peculiarly required when a continual production of such matters is taking place through the disintegration of the nervous and muscular tissues consequent upon their functional activity, and which is already retarded by the diminution in the activity of respiration. We shall presently find that experience is here also in accordance with theory; the result of many trials having shown, that severe and long-continued

---

Depuration — purification.

[2] *Excrementitious* — worn-out, refuse, deleterious matters, which, unless regularly removed from the blood, impair its purity

[3] *Insensible transpiration* — the continued discharge of vapour by the skin in an imperceptible form.

[4] We are of course supposing throughout, that water is freely supplied in both cases. The exhaustion produced by the undue diminution of the fluids of the body, indicated by excessive thirst, is of quite a different character.

[5] *Elimination* — separation; removal.

10 *

exertion in tropical climates can be better sustained *without* alcoholic liquors, than *with* their habitual use.

135. The stimulative effects, from which Alcoholic liquors derive their reputation as supporters of bodily vigour, during habitual exposure to a heated atmosphere, are exerted in two ways; in the first place, by giving temporary assistance to the digestive process; and secondly, by increasing, for a time, the nervous and muscular power. It is commonly supposed that the diminution of appetite, which is experienced by most persons who change their residence from a temperate country to a hot one, is the result of the enervating influence of the climate; whereas the fact is evident to those who take into account the proportionally smaller amount of carbonic acid exhaled as the external temperature rises, that the diminished appetite chiefly results from diminution in the demand for combustive material; and that it ought, therefore, to be taken as an indication of the propriety of lessening the amount of food ingested, rather than of forcing the stomach to augmented activity for the purpose of disposing of the superfluity which it has taken in. All medical authorities on the diseases of tropical climates are in accord upon this point,—that, next to the injury derived from the abuse of fermented liquors, excess in diet is one of the most fertile of those sources of disease which arise out of the personal habits of the individual; and such excess is in great degree due to the use of alcoholic stimulants as an artificial provocative to the appetite, whereby the blood becomes charged with more alimentary material than it can rightly dispose of; so that the diminution in the activity of the respiratory process throws the elimination of this superfluity upon the liver, which organ consequently becomes peculiarly liable to functional disorder.

136. We have continual opportunities of noticing the same sequence of phenomena in our own country, though in a less marked degree. A diminution in the appetite is experienced by most persons during the heat of summer; and if the warning be not lost, the amount of food ingested is proportioned to the demand. But those who from habit continue to take in their usual supply, are extremely liable to be warned of the impropriety of such a course, by hepatic derangement; and the bilious diarrhœa which is so common in the latter part of summer, and which is connected in the popular mind with the "plum season," (although it frequently affects persons who have altogether abstained from fruit,) seems to find a rational explanation in the accumulation of excrementitious matter, which must be the consequence of habitual excess in diet,—especially when the stomach is stimulated by alcoholic liquors to digest more than could be appropriated without such artificial aid.

137. There is no reason whatever to believe that (with the exception of the difference in regard to amount which has been already

remarked upon, § 129,) the stimulating influence of Alcoholic liquors upon the nervous system, whereby it is enabled to put forth increased power so long as this influence lasts, is exerted in any other mode, when the body is habitually exposed to a high temperature, than that in which it operates under ordinary circumstances. That the excitement must be followed by subsequent depression, is as true in India, as in England ; and that this excitement, if habitually had recourse to, will be followed in hot climates by consequences even more injurious than in cold or temperate regions, might be inferred from all that has been already stated in regard to its peculiar unsuitableness when the activity of the respiratory process is diminished.

138. We shall now proceed to inquire, therefore, how far the experience, both of individuals and of large bodies of men, supports the idea, that Abstinence from alcoholic stimulants, or at most the very sparing use of them, is favourable to the endurance of extreme heat, especially when great bodily exertion is required. And we shall first cite the evidence of the late Mr. Gardiner,[1] a well educated surgeon, who spent several years of most active exertion in the exploration of the Botany of Brazil, into which country he penetrated further than any scientific European had previously done. During three years' travelling in that climate, he tells us,[2] under constant fatigue and exposure to vicissitudes of weather and irregularity of living, his only beverage, besides water, was tea, of which he had laid in a large stock previously to his departure from Pernambuco. He was told when he arrived at Brazil, that he would find it necessary to mix either wine or brandy with the water which he drank ; but a very short experience convinced him, not only that they are unnecessary, but that they are decidedly hurtful to those whose occupations lead them much into the sun. " Whoever drinks stimulating liquors," he says, " and travels day after day in the sun, will certainly suffer from headache ; and in countries where miasmata prevail, he will be far more likely to be attacked by the diseases which are there endemic."

139. Equally explicit testimony is borne by Sir James Brooke, the enterprising and skilful colonizer of Borneo ; who speaks in his " Journal " of habitual abstinence from alcoholic liquors as decidedly conducive to the maintenance of health, and of the power of sustained exertion, in the equatorial regions in which he had established himself.—So, again, Mr. Waterton, the well-known traveller, speaks of himself as confident that the preservation of his vigour during

---

[1] The writer has been informed by an intimate friend of this gentleman, that his lamented death, which took place from a *coup de soleil* (sun-stroke), whilst holding the appointment of Superintendent of the Botanic Garden at Ceylon, was entirely due to the injudicious and almost fool-hardy exposure to which his confidence in his vigour led him to subject himself.

[2] Travels in Brazil, p. 402.

many years of toil and exposure in tropical climates, is mainly due to his total abstinence from fermented liquors.—And the writer has been assured by Dr. Daniell, who was for a long time stationed as medical officer in the equatorial portions of Western Africa, that he found the use of the ordinary alcoholic liquors decidedly inimical to the power of exertion; the strongest beverage which could be habitually made use of without injury, being the 'palm-wine' of those countries, which is very little, if at all, more alcoholic than our ginger-beer. The following testimony, given by Doctor Mosely in his work on Tropical Diseases, may be added to the foregoing:— " I have ever found," he says, " from my own knowledge and custom, as well as from the custom and observation of others, that those who drink nothing but water, or make it their principal drink, are but little affected by the climate, and can undergo the greatest fatigue without inconvenience." Many other individual testimonies might be cited to the same effect; but as these are open to the objection of being influenced by peculiarities of individual constitution, it will be preferable to have recourse to cases in which large bodies of men are included.

140. The following statement, which the writer has received from an Officer in the regiment to which it refers, proves that our English soldiers in India not only do not suffer from, but are absolutely benefited by, abstinence from Alcoholic liquors during a continuance of unusually severe exertion. "In the early part of the year 1847, the 84th Regiment marched by wings from Madras to Secunderabad, a distance of between four and five hundred miles. They were forty-seven days on the road, and during this period the men were practically speaking teetotalers. Previously to leaving Madras, subscriptions were made among the men, and a coffee establishment was organized. Every morning, when the tents were struck, a pint of hot coffee and a biscuit were ready for each man, instead of the daily morning dram which soldiers on the march in India almost invariably take. Half way on the day's march, the regiment halted, and another pint of coffee was ready for any man who wished it. The regimental canteen was opened only at ten and twelve o'clock for a short time, but the men did not frequent it; and the daily consumption of arrack for our wing was only two gallons and a few drams per diem, instead of twenty-seven gallons, which was the daily Government allowance. The commanding officer employed the most judicious precautions to prevent the men from obtaining arrack in the villages on the route; and his exertions were effectively seconded by the zealous co-operation of the other officers, and by the admirable conduct of the majority of the men, who were fully persuaded of the obnoxious influence of ardent spirits during exercise in the sun. The results of this water-system were shortly these:—Although the road is proverbial for cholera

and dysentery, and passes through several unhealthy and marshy districts, the men were free from sickness to an extent absolutely unprecedented in our marches in India; they had no cholera and no fever; and only two men were lost by dysentery, both of whom were old chronic cases taken out of hospital at Madras. With these exceptions, there was scarcely a serious case of sickness during the whole march. The officers were surprised that the men marched infinitely better, with less fatigue and with fewer stragglers, than they had ever before known; and it was noticed by every one that the men were unusually cheerful and contented. During the whole march, the regiment had not a single prisoner for drunkenness." A considerable proportion of the men (the writer has learned from his informant) abstained entirely from arrack; and the consumption of those who occasionally took it, was far below their usual allowance. Those who *entirely* abstained were certainly in no respect inferior, either in power of sustaining exertion, or in freedom from sickness, to those who occasionally took small quantities of spirits; on the contrary, they rather seemed to have the advantage. That this remarkable result was not due to any peculiar healthfulness of the season, or other modifying circumstance, is shown by the fact that the 63rd Regiment, which performed *the same march, at the very same time*, though in the opposite direction, lost several men out of a strength of 400; and that it had so many sick, that when it met the 84th on its march, it was obliged to borrow the spare "dhoolies" (or palanquins for the sick) belonging to the latter.[1]

141. The foregoing account fully accords with that given by Sir James (then Mr.) McGrigor, of the march in Egypt of a division of the British army sent from Hindustan to aid the main army in opposing the French under Buonaparte. After the Great Desert had been crossed, in July, 1801, no spirits were issued to the troops in Upper Egypt, owing to a difficulty in procuring carriage for them. At this time there was much fatigue-duty to be performed; which, for want of followers, was done by the soldiers themselves; the other duties were severe upon them; they were frequently exercised, and were much in the sun; the heat was excessive, tho thermometer standing at 113° or 114° Fahr. in the soldiers' tents in the middle of the day; *but at no time was the Indian army more healthy.*[2]

---

[1] The marked contrast between the rate of mortality in the 63rd and 84th Regiments, during their respective residences at Secunderabad during two consecutive years, has been already noticed (§§ 81, 82); but it may be as well here to remind the reader, that the former lost 73 men in nine months, which was at the rate of 78·8 per 1000 of average strength for the entire year; whilst the latter lost but 39 men in the whole twelve months, being at the rate of 34·2 of average strength.

[2] Medical Sketches of the Expedition from India to Egypt, p. 86.

142. The intimate acquaintance of Sir Charles Napier with the habits and wants of the Indian soldier can be doubted by no one; and the following is his testimony in favour of the abstinence system, (delivered in his own characteristic manner) as contained in his address to the 96th Regiment, when he reviewed it at Calcutta on the 11th of May, 1849.—"Let me give you a bit of advice—that is, don't drink. I know young men do not think much about advice from old men. They put their tongue in their cheek, and think that they know a good deal better than the old cove that is giving them advice. But let me tell you that you are come to a country where, if you drink, you're dead men. If you be sober and steady, you'll get on well; but if you drink, you're done for. You will be either invalided or die. I knew two regiments in this country, one drank, the other didn't drink. The one that didn't drink is one of the finest regiments, and has got on as well as any regiment in existence. The one that did drink has been all but destroyed. For any regiment for which I have a respect (and there is not one of the British regiments that I don't respect) I should always try and persuade them to keep from drinking. I know there are some men who will drink in spite of the devil and their officers; but such men will soon be in hospital, and very few that go in, in this country, ever come out again."

143. Whatever temporary advantage, then, is derived or supposed to be derived from the stimulating powers of Alcoholic liquors, when they are used with a view of sustaining the power of exertion in tropical climates, is dearly purchased by the increased liability to disease, which not only *theoretically*, but according to all competent evidence, *actually* results from their habitual use. And thus Theory and Practice are again completely agreed, in affording a decisive contradiction to the usually received idea, that Alcoholic liquors assist the body in the endurance of Heat.

V. — RESISTANCE TO MORBIFIC AGENCIES.[1]

144. It is a common idea, and one apparently supported by adequate evidence, that such a use of fermented liquors as aids in keeping the body in "high condition," renders it less susceptible of the influence of pestilential miasmata,[2] of cold and damp, or of other morbific agencies; and this belief is entertained by many, who deprecate the habitual use of fermented liquors under other circumstances. Thus, says Dr. Macnish, "I am persuaded that while, in the tropics, stimulating liquors are highly prejudicial, and often occasion, while they never prevent disease; they are fre-

---

[1] *Morbific agencies* — disease-producing agencies.
[2] *Miasmata* — the morbific, or disease-producing emanations from domestic and personal impurities, decaying organic substances, marshes, &c.

quently of great service in accomplishing the latter object in damp foggy countries, especially when fatigue, poor diet, agues, dysenteries, and other diseases of debility are to be contended against." — "In countries subject to intermittents, it is very well known that those who indulge moderately in spirits are much less subject to these diseases than the strictly abstinent." [1] These assertions he endeavours to justify by the two following statements. "At Walcheren it was remarked that those officers and soldiers who took schnaps, *alias* brandy drams, in the morning, and smoked, escaped the fever which was so destructive to the British troops; and the natives generally insisted upon doing so before going out in the morning." [2] Again, "A British regiment quartered on the Niagara frontier of Upper Canada, in the year 1813, was prevented by some accident from receiving the usual supply of spirits; and in a very short time, more than two-thirds of the men were on the sick list from ague and dysentery; while the very next year, on the same ground, and in almost every respect under the same circumstances, except that the men had their usual allowance of spirits, the sickness was extremely trifling. Every person acquainted with the circumstances, believed that the diminution of the sick during the latter period, was attributable to the men having received the quantity of spirits to which they had been habituated." [3]

145. Now it is obvious that neither of these facts proves that exposure to the morbific agencies in question renders an allowance of spirits necessary, or even beneficial, for those who have *not* been accustomed to make use of it under ordinary circumstances. On the contrary, the second instance is a valuable testimony to the disadvantage of habitual dependence upon alcoholic stimulants; inasmuch as it is evident that, when they were withheld from the troops, the constitution of the men was rendered peculiarly susceptible to the causes of disease indigenous [4] to their locality. All that it proves is, that an unduly depressed state of the system is favourable to attacks of ague and dysentery (of which every medical practitioner is aware), and that in persons who have habituated themselves to the use of spirits, such depression is liable to supervene when the allowance is withheld, and may be for a time kept off by its restoration. — And even the first example cannot be said to prove more than this; for it simply gives us the experience of individuals who took an early dose of spirits, as compared with that of the individuals who abstained from this habit; without

---

[1] Anatomy of Drunkenness, pp. 277, 279.
[2] Glasgow Medical Journal, No. xv.
[3] Op. cit.
[4] *Indigenous* — belonging to the place.

telling us that the latter adopted any of those substitutes, which prudential experience would dictate.

146. The writer is strongly impressed with the belief, that the result, in this and in many similar cases, is to be attributed to the neglect of such precautions. It is well known that in localities where zymotic [1] poisons are indigenous, no condition of the healthy system is so obnoxious to their influence, as that which is natural to it on first rising in the morning, when the stomach is empty, the pulse comparatively feeble, and the heat-producing power nearly at its minimum. The nutritive actions which have been taking place during repose, have prepared the nervous and muscular apparatus for renewed activity; but this has been accomplished at the expense of the blood, from which there has been a continual drain, both for the regeneration of the tissues, and for the maintenance of the animal heat. It is within the experience of most persons, that nervous and muscular exertion are less efficiently sustained,[2] and external cold less fully resisted, at this period, than at any other; and the recommendation of experience to "take something to keep the cold out of the stomach" is here fully justified upon physiological principles. But it does not hence follow, that alcoholic stimulants constitute the best means of protecting the system against the influence of morbific agencies; on the contrary, we shall find strong reason to believe that other means, properly employed, would be as efficacious at the time, and would have a more permanently beneficial effect.

147. A man previously in the enjoyment of vigorous health, and not accustomed to depend upon alcoholic stimulants, will derive all the protection he can require, from taking his first solid meal before he exposes himself to the cold, damp, or pestilential miasmata, whose influence is to be resisted; and the moderate use of hot tea, coffee, or cocoa, will help to diffuse a genial warmth through his body, which is more enduring than that which results from the ingestion of spirituous liquors. In this way the stomach will be wholesomely employed, new material will be supplied to the blood, the circulation will be quickened without being excited, the firmness of the pulse will be increased, and the heat-producing power will be augmented; and all this, in a manner strictly accordant with the normal economy of the bodily system.—On the other hand, although the use of spirits, by producing a temporary excitement of the circulation, will probably render the system less

---

[1] *Zymotic*—fermentive. A term applied to the causes productive of endemic, epidemic, and contagious diseases.

[2] The writer can speak feelingly on this subject; being himself unable to walk a couple of miles upon the empty stomach of early morning, without extreme languor almost amounting to syncope; although four or five times that amount would usually be performed by him after breakfast with scarcely an approach to fatigue.

obnoxious to morbific agencies, than if it be exposed to them before its dormant energies have been in any way aroused, yet we can scarcely anticipate that they can be as favourable to the *sustenance* of its energy (a previously healthy and vigorous condition being supposed), as persistence in the regular habits to which it has been accustomed. For it has been already shown, that the continued endurance of cold is not favoured by the use of alcoholic liquors, but on the contrary is impaired by it; and where cold, therefore, acts concurrently with zymotic poisons, and favours their operation by the depression of the vital powers which it induces, we should feel certain that those means would be most conducive to the resisting power of the system, which are most efficient in maintaining its standard warmth.

148. So far as we are acquainted with the bearings of experience on this question, they are decidedly favourable to the view here advocated, namely,—that where a healthy state of the system has been previously maintained without the assistance of alcoholic liquors, the operation of morbific agents will be more efficiently warded off by a continuance of the abstinent plan, than by recourse to stimulants; provided that the same precaution be exercised by the disciple of abstinence, as by the spirit-drinker, in not exposing himself to the morning air without a fortification of "the inner man." For we do not see that the circumstances of tropical or those of cold or temperate climates differ, as regards the susceptibility of the system to zymotic poisons, in any other particular than their temperature; but this will act in more than one way; for whilst, on the one hand, the depressing influence of cold upon the body will tend to increase its susceptibility, the agency of heat, on the other, will augment the potency of the poison. Now, that abstinence from spirits diminishes instead of increasing the liability of the body to the influence of pestilential miasms in warm climates, provided that other precautions be duly taken, we have not merely the individual experience of Mr. Gardner, Mr. Waterton, and others (§§ 138, 139), in addition to the testimony of many medical observers, but the important evidence derived from the march of the 84th Regiment along a road "proverbial for cholera and dysentery," without a single fresh attack of these complaints (§ 140). Here the "pint of hot coffee and biscuit" were certainly to the full as efficacious as "the daily morning dram, which soldiers on the march in India almost invariably take;" and there is no adequate reason why the coffee should not have an equal value in colder countries, when employed with the larger allowance of heat-producing food which will be there required.

149. When the *remote* effects of the two systems are contrasted, there can be little hesitation in assigning the preference to the abstinent plan. For the object being to sustain the utmost *equa-*

11

bility of health, and especially to avoid that depressed condition which sooner or later supervenes upon states of undue excitement, it is obvious that when all the nutritive functions are regularly and vigorously discharged, it is unwise to interfere with their performance by the use of Alcoholic liquors, which, if sufficient to produce either general stimulation, or excitement of any one function, must involve as its consequence a corresponding diminution of the normal activity at some subsequent period. It is quite true, that this may not manifest itself at once; so that for weeks, months, and years, the vigour of the system may seem to be efficiently maintained, and morbific agencies to be perfectly kept at bay, by the habitual use of a small quantity of alcoholic stimulus; to which the beneficial result will then be probably attributed. But the trial is not complete in weeks, months, or years; it must last for the whole life; and if it be true, as we shall presently endeavour to show, that the continued employment, however moderate, of small quantities of alcoholic liquors, favours, if it does not necessarily induce, an early exhaustion of the vital powers, it cannot be questioned that the system will then be left in a state of peculiar susceptibility to the influence of zymotic poisons and other morbific agencies. It is well known that persons of regular habits and good ordinary health, who have long resided in countries where intermittent fevers prevail, are frequently attacked by them when their vital powers begin to decline with advancing years; and if that decline be hastened by the previous over-excitement of alcoholic liquors, the influence of these morbific causes will be earlier and more powerfully exerted.

150. These theoretical deductions are not merely sanctioned by such results of experience as can be brought to bear directly upon them; for they are in complete harmony with the facts universally admitted, in regard to the peculiar susceptibility of habitually intemperate persons, and especially of those whose constitutions have been broken down by the combined influence of intemperance and advancing years, to the attacks of fever, cholera, and other pestilential disorders (§ 65). For, we again repeat, if the cause, when acting with its greatest potency, is attended with a result which no one can hesitate in accepting, it is but reasonable to attribute to the same cause, acting with diminished intensity, but over a longer period of time, a result of a similar nature; even though this may be so long postponed, that its dependence on that cause is in danger of being overlooked.

151. We have abundant evidence, then, not merely in the experience of individuals, but in that of large bodies of men, that the most vigorous health *may be* maintained, under circumstances usually regarded as most trying to the power of bodily and mental endurance, without the assistance of Alcoholic stimulants. Such evidence is afforded by the numerous ships that are travelling every

part of the wide ocean, whose crews, pledged to the total abstinence principle, maintain a degree of health and vigour which cannot be surpassed; by the many workshops of every kind, in which the severest labour is endured, with a constancy at least equal to that of the drinkers of alcoholic beverages; by troops executing toilsome marches in the sultry heat of a torrid zone, and through the pestilential atmosphere of tropical marshes, who find the " cup of cold water " more refreshing and sustaining than the spirituous drinks which hurry so many of their comrades to an early grave; and by numbers of men and women, in every rank of life, in every variety of condition, and subjected to every kind of mental and bodily exertion, who have given the principle of Total Abstinence a fair trial, and have borne their willing testimony to its beneficial results. And where such is the case, there can scarcely be a question that this system is preferable to the habitual use, however moderate, of fermented liquors. For, if the appetite prompt to the use of an adequate amount of nourishment to repair the waste of the sytem ; if the stomach perform its action with due energy, and supply to the absorbent vessels the material for fresh blood in a state of due preparation ; if the circulation be carried on with that equable regularity, which is most favourable to the actions to which it is subservient; if the various tissues draw from the current of nutritious fluid the materials which they severally require, and apply these materials to their own maintenance and regeneration ; if the lungs freely exhale the carbonic acid which is evolved by their exercise, and introduce the oxygen which is needed for a renewal of the effort; and if the liver, kidneys, and skin, by the constant discharge of their respective offices, eliminate from the blood the other products of the waste of the system, and thus keep it in the state of purity most favourable to the discharge of its multitudinous functions;—in a word, if all the actions concerned in the maintenance of the fabric be already discharged with that vigour and uniformity which constitute *health*, why should we attempt to alter them by means of agents, which, if they produce any effect whatever on the system, can only operate by producing a departure from that perfect balance of the several parts of the nutritive functions which it is so desirable to maintain, and so difficult to restore when perverted? Let us examine these questions in more detail.

VI.—CONSEQUENCES OF THE HABITUAL "MODERATE" USE OF ALCOHOLIC LIQUORS.

152. *Effect upon the General System and Excretory Organs.*— If the natural appetite be already good enough to give a relish to the food which the system requires, can the artificial production of an increased appetite be necessary or desirable? And if the sto·

mach be already capable of digesting and preparing as much nutriment as is required to keep up the solids of the blood to their proper amount, can any but prejudicial consequences result from forcing it to dispose of more? Two classes of evils may be expected to proceed from such a system; in the first place, the habitual introduction of more alimentary material into the circulating current than the nutritive functions can appropriate, must predispose to disorder of the system in general; and secondly, by constant reliance upon an artificial stimulus, the natural powers of the stomach itself must be in danger of becoming gradually impaired.

153. The effect upon the system at large, of an habitual introduction of more alimentary material than the nutritive functions can appropriate, seems to vary with the temperament. In some individuals, they are converted into blood, so that the normal amount of that liquid undergoes an augmentation; thereby inducing a state of *plethora*,[1] which is favourable to local congestions and inflammatory diseases of various organs, and which especially predisposes to hemorrhage — this being an effort of nature to relieve the undue turgescence. But in other constitutions, the superfluous aliment would seem to be never so far vitalized and assimilated, but is from the first destined to excretion; the lungs, the liver, the kidney, and the skin, are thus called upon to remove, not merely the products of the normal waste or disintegration of the system, but also the superfluous non-assimilated matter; and hence they are brought into a state of undue functional activity, which cannot but render them peculiarly susceptible of derangement. The excretory action of the lungs, however, is chiefly regulated (as already shown, § 127), by the temperature; so that, when it is diminished by external warmth, more remains to be accomplished by the other depurating[2] organs; and hence any excess in diet is more likely to have a prejudicial effect upon the latter in warm climates, and during the summer, than in a colder atmosphere.

154. This is precisely what experience teaches. From habitual excess in diet, in temperate climates, in persons not of the sanguineous temperament, disorders both of the Liver and Kidney are very apt to arise; those being most liable to the former, who have not the power of generating fatty tissue at the expense of the surplus of non-azotized food; and those being most liable to the latter, in whom the too free use of alcoholic liquors occasions an undue determination of blood to the Kidneys. On the other hand, habitual excess of food in warm climates usually manifests itself first in disorders of the Liver; since the diminished excretion of carbon by the lungs causes the blood to proceed to the liver more highly charged

---

[1] *Plethora* — overfulness of blood.
[2] *Depurating* — purifying.

with that element, whilst at the same time the consumption of that part of the biliary secretion which should be normally oxygenated and carried off through the lungs, is interfered with. On the other hand, the Skin, whose functions are greatly increased in activity, comes to the assistance of the Kidneys in disposing of the superfluity of azotized[1] aliment; a considerable amount of *urea*[2] being daily excreted through the former channel.[3] This result of what is accounted the "moderate" use of alcoholic liquors in warm climates, for the purpose of increasing the appetite and stimulating the digestive powers of the stomach, is much dwelt upon by writers on tropical diseases; who represent it as, in the long run, not less hurtful than that excess which produces effects more immediately and obviously pernicious. In this point of view it ranks with high-seasoned dishes, and those other seducing provocatives to the diminished appetite and lessened digestive powers of the residents in such climates, which, by occasioning the habitual ingestion of more food than the system requires, are among the most fertile sources of tropical disease.

155. Now, as already remarked, almost every cause of disease acts on the human system with greater potency in tropical than in temperate regions; and we have opportunities, therefore, in the study of tropical diseases, of perceiving the agency of causes, whose tardiness of operation under other circumstances interferes with our recognition of their real results. It cannot, then, be imagined that even a small habitual excess in diet, induced by the stimulating action of fermented liquors, can be without its remote consequences upon the general system; even though it may be for a time sufficiently compensated by increased activity of the excreting organs.[4] And the disorders of the Liver and Kidneys, which are so frequent among those who have been accustomed to this mode of living for many years, without (as they believe) any injurious consequences, are, as surely to be set down to it, as are those congestive and inflammatory diseases of the abdominal viscera, which so much more speedily follow upon habitual excess in warm climates. For the excreting organs cannot be always kept in a condition of excessive activity; like other parts of the system, they suffer sooner or later from too great an exaltation of their function; and if this should not pass, as it often does, into an inflammatory condition, it is almost certain to be remotely followed by a state of depressed activity, in which the nutrition of the organ becomes impaired, so that it is left,

---

[1] *Azotized aliment* — aliment containing nitrogen.
[2] *Urea* — a substance forming an essential part of the urine.
[3] See the experiments of Dr. Landerer of Athens, in Brit. and For. Med. Chir. Review, vol. i. p. 541.
[4] *Excreting organs* — organs that separate the refuse, worn-out, and deleterious matters from the blood.

during the remainder of life, in a state by no means equal to the performance of its regular duties.

156. In asserting that to the ordinary use of fermented liquors in "moderate" quantity, during early and middle life, and to the habitual excess in diet (however slight) to which they prompt, we are to attribute many of the chronic disorders of the excreting organs which are amongst the most common ailments of advancing years, we may seem to go beyond the positive teachings of experience. The consequences are so remote, that we may not appear to be justified in attributing them to the causes we have assigned. But let it be remembered that we have multitudes of other cases, in which the long-continued agency of morbific causes of comparatively low intensity has been proved to be in the end not less potent, than the administration of a poison in a dose large enough to produce its obviously and immediately-injurious effects. Thus, a man who would be rapidly suffocated by immersion in an atmosphere of carbonic acid, may live for weeks, months, or years, in an atmosphere slightly contaminated by it, without experiencing any evil effects which he can distinctly connect with its influence ; and yet who will now deny that the constant action of this minute dose of aerial poison is insidiously undermining his vital powers, and preparing him to become the easy prey of the destroying pestilence ? So again, we see that a brief exposure to the pestilential atmosphere of the swamps of the Guinea coast, is often sufficient to induce an attack of the most rapidly fatal forms of tropical fever ; but the dweller among the marshy lands of temperate climates, inhaling the paludal poison [1] in its less concentrated form, becomes after a time afflicted with intermittent fever ; and no one has any hesitation in here recognizing the connexion of cause and effect. On the other hand, the resident in a town, where the insufficiency of the drainage causes the surface-moisture to be imperfectly carried off, and to be not merely charged with the malaria of vegetable decomposition, but with the miasmatic emanations of animal putrescence, may long be free from serious disorder, if the cause do not operate in sufficient intensity ; yet he becomes liable in a greatly-increased degree to the operation of almost every morbific agent, and especially of the various forms of fever-poison ; and no one who has paid even a slight degree of attention to the results of the sanitary enquiries which have now been carried on for many years past, hesitates in admitting the relation of cause and effect between insufficiency of drainage and the higher rate of mortality in undrained localities, although not only days and weeks, but months and years, may be required for the operation of the cause upon the animal system.

157. Should we not then be running counter to all analogy, if

---

[1] *Paludal poisons* — exhalations from marshes

we do not hold ourselves ready to admit, that such an habitual excess in diet as is favoured by the moderate use of Alcoholic stimulants, and a consequent habitual over-exertion of the excretory organs, must be likely to have remotely injurious results; and are we not justified in assuming a relation of cause and effect to exist, when we find such results occurring precisely as we should predict? If the medical man has no hesitation in regarding those severer derangements of the excretory organs, which are so common amongst those who commit habitual excesses in eating and drinking, as the consequence of those excesses, why should he refrain from attributing the milder but more protracted disorders of the same organs, to the less violent but more enduring operation of the same cause?—"The little I take, does me no harm," is the common defence of those who are indisposed to abandon an agreeable habit, and who cannot plead a positive benefit derived from it; but before such a statement can be justified, the individual who makes it ought to be endowed with the gift of prophecy, and to be able to have present in his mind the whole future history of his bodily fabric, and to show that by reducing the amount of his excess to a measure which produces no immediately-injurious results, he has not merely postponed its evil consequences to a remote period, but has kept himself free from them altogether. The *onus probandi*[1] lies with those who assume the *absence* of a connection, which is indicated by every fact with which we are acquainted.

158. Although we have hitherto been considering the effects of the "moderate" use of Alcoholic stimulants upon the excretory organs, as consisting simply in augmenting the amount of labour they are called upon to perform, by favouring the reception of too large an amount of alimentary matter into the system, yet there is another point of view under which it will be convenient here to examine its results; namely, the direct influence of the alcoholic stimulus upon the organs themselves. This influence may for a time be corrective of the other, and may thus aid in concealing and retarding its evil consequences. For we have seen (§ 57) that the introduction of a small quantity of alcohol into the circulation has a direct action upon the Kidney, increasing the determination of blood to that organ, and tending to augment its secretion; and it is highly probable that it has a similar effect upon the Liver, more especially as the blood which has received the alcohol by the absorbent action of the gastric veins,[2] passes through that organ before proceeding to any other part of the system. In this manner, the call for increased action of these two depurating organs[3] being met by augmented

---

[1] *Onus probandi* — burden of proof; the duty of proving.

[2] *Gastric veins* — veins of the stomach.

[3] *Depurating organs* — organs that remove refuse, worn-out, and deleterious matters from the blood.

functional activity on their part, a system of compensation is main-tained, whereby the effects of excess are neutralized for a time,—but *only* for a time; for as surely as any organ is habitually exerted in an excessive degree, so surely must its vital powers be prematurely exhausted, the remoteness of the period at which the flagging of its power begins to manifest itself, being inversely to the degree of habitual over-excitement. Hence we have additional reason for imputing a considerable proportion of those chronic disorders of the excretory organs to which reference has been more especially made, to the habitual employment of alcoholic liquors, in what is ordinarily considered to be a "moderate" amount, and regarded as perfectly consistent with health, if not required to maintain it.

159. It would be absurd, however, to affirm that such diseases *always* proceed from this cause; since those who practise Total Ab-stinence from alcoholic liquors are by no means proof against other errors in dietetics; and in so far as they habitually take in more food than their system needs, they will be liable to suffer from dis-order of the organs whose duty it is to eliminate the waste. But they will be much sooner warned of the excess they have com-mitted, if the stomach refuses to digest the superfluity, instead of being forced by artificial stimulation to an undue exertion of its power; and an attack of indigestion, by early giving a salutary check to the practice, may ward off its remoter consequences. It is the freedom from such checks, up to a certain time of life, which encourages in those, who habitually use fermented liquors in "mo-deration," and who at the same time practise habitual though slight excess in the amount of solid food which they consume, the delu-sive belief that in neither case are they doing themselves any harm.

160. *Effect upon the Stomach.*—Such, then, are the consequences to the system at large, which Theory and Experience join to indi-cate, as resulting from such an habitual use of Alcoholic liquors as stimulates the appetite to desire, and the stomach to digest, a larger amount of food than is necessary to supply the wants of the body; and we have next to inquire into the effects it produces upon the Stomach itself. We have already described the admitted results of what is commonly regarded as "excess;" and we shall therefore at present limit ourselves to the inquiry, whether the "moderate" use of alcoholic liquors is likely to be productive of any injurious consequences, as regards this important organ. All our knowledge of the action of stimulants would lead to the conclusion, that when once the habit of employing them has been established, and the stomach is come to rely (as it were) upon the extraneous aid which they afford, its power of performing its duty without such aid must be impaired. The case is very similar to that of sleep. A person in health, and not subjected to any unfavourable influences, is natu-

rally disposed to pass as much time in repose, as his system needs
for its renovation ; but if he were long to accustom himself to the
use of a narcotic, he would find himself completely unable to sleep
without it. And experience shows, in like manner, that those who
have long been habituated to the moderate use of Alcoholic beverages with their meals, are seldom able to discontinue them without
a temporary loss of appetite and of digestive power,—unless, indeed, their place be supplied by the more wholesome excitement of
fresh air and exercise.

161. With many persons, the evil, so far as the Stomach is concerned, may seem to be confined to the induction of this state of
reliance on artificial aid. Year after year passes away, without any
indication that its powers have been overtasked, or that any unhealthy change has taken place in its circulation or nutrition ; and
the usual dose of the alcoholic stimulant appears still to produce its
wonted effect. But this does not show that the practice is really
innocuous. We have seen that whilst a potent dose of a poison
speedily manifests its action by the violence of its effects, the repeated introduction of minute doses is not really inoperative, although the effects are not speedily apparent. If the stomach be
not an exception to the general law of the action of stimulants upon
the animal body, we should expect that by the habitual over-excitement of its function, in however trifling a degree, its vital energy
will undergo a premature depression ; and that the result of the
moderate use of alcoholic stimulants will manifest itself, sooner or
later, in diminution of the digestive power. The earliest indication
of this, in most instances, is the demand for the augmentation of
the stimulus to produce the same result ; the amount which was at
first sufficient to whet the appetite and increase the digestive power,
being no longer found adequate. If the demand be yielded to, and
the quantity of the stimulus be augmented, the original benefit
seems for a time to be derived from it ; but after the stomach has
become tolerant of the liquor, that which at first excited it to increased functional activity, does so no longer, and a further increase
is called for ; until what began in "moderation" ends in positive
excess, with all its consequent evils. But supposing this demand
not to be felt, or not to be yielded to, the same "moderate" allowance being indulged in for a long course of years, we should antici
pate that injurious consequences, though perhaps long postponed,
must ultimately show themselves ; and that such is the case, is unfortunately the experience of a vast number, who suffer, by that
"loss of tone" of the stomach which is so common an attendant
of advancing years, for the too great activity to which the organ
has been previously forced, during the long period of early and
middle life. And although the common idea, that alcoholic liquors

when taken in small quantities have a *tonic*[1] property, may render it difficult for some to coincide in the conclusion that the real effect of the habitual use of even this small quantity must be of the opposite kind,—exhaustive instead of tonic,—yet as this idea has no other foundation than the temporary assistance derived from the stimulating powers of alcohol, it ought not to prevent our recognition of the consequences which might be theoretically expected to proceed from its long-continued action.

162. It is not here maintained, however, that the habitual employment of alcoholic stimulants in small quantity, even when coupled with habitual excess in the amount of food ingested, uniformly stands to the loss of appetite and digestive power so frequent with the advance of years, in the relation of cause and effect; for there can be no doubt that the habit may be persevered in by some individuals throughout a long life, without the manifestation of any injurious results; whilst on the other hand, it cannot but be admitted that the disorder in question may be induced in other ways. But the existence of exceptional cases by no means invalidates the argument based upon general experience; any more than our occasionally meeting with individuals who have daily consumed a bottle of spirits, and have yet enjoyed a hearty old age, warrants us in rejecting the evidence which indicates that such a consumption would have, in *by far the* larger proportion of mankind, a decided tendency to shorten life. Nor does it follow, that because the loss of digestive power may be justly attributed to other causes when this one has been wanting, it has been inoperative when present. In fact, there can be little doubt that amongst the class of men who are engaged in active mental occupation, and who justify a moderate use of alcoholic liquors on the ground that it keeps them "up to their work," the expenditure of nervous power, consequent upon that undue exertion of the cerebral functions which has been aided by the continual over-stimulation, has a large share in the result.

163. *Effect upon the Nervous System.* — Every medical man is familiar with cases, in which the "wear and tear" of an over-active life has been sustained with little apparent loss of power for perhaps a long series of years; but in which there is a sudden failure both of mental and bodily vigour, as manifested in deficiency of power of continued mental exertion, depression of spirits, want of appetite, enfeebled digestion, and the whole train of disordered actions which is consequent upon this condition. It is not to be denied that such a state may arise quite independently of the agency, direct or indirect, of habitual stimulation; one instance, in particular, is strongly present to the writer's recollection, in which it supervened on a long course of excessive mental exertion, in an individual who was most

---

[1] *Tonic* — imparting permanent strength and vigour.

moderate in everything but the labour of his brain, and who rarely or never sought for artificial support from alcoholic stimulants. But the most common case is that, in which two sets of causes are in action together. An habitual system of over-exertion of the nervous system may be maintained for a longer time by many persons, with the assistance of alcoholic stimulants, than without them; and thus the delusion is kept up, that the strength is not really overtasked: when the fact is, on the contrary, that the prolongation of the term of over-exertion, by the repeated application of the stimulus, is really expending more and more of the powers of the nervous system, and preparing for a more complete prostration at a later period.

164. The temporary advantage, then, which is thus gained, is very dearly purchased. The man who habitually abstains, not merely from alcoholic liquors, but from other artificial provocatives, (misnamed supports,) to the endurance of mental activity, is early warned by the failure of his intellectual energy and cheerful tone of spirits, that he is over-tasking his brain; whilst his stomach tells the same tale in another way, — the failure of power to digest that which the fabric really needs for its regeneration, being indicative of an exhaustion of nervous energy. A short period of rest and change, in such a condition, is usually sufficient for the renovation of the system, and for the recovery of the mental and bodily vigour. But the case is very different, when the effort has been sustained for a lengthened series of years, by means of the delusive support afforded by alcoholic liquors; for as the excessive expenditure of nervous power has been greater, so is the exhaustion more complete; and as the stomach has been longer over-excited and over-tasked, its tone is the more seriously injured, not merely by the depression consequent upon its own over-work, but by the impairment of the nervous power which is required for its due activity. Thus, then, although the consequences of habitual over-exertion of the brain may be less speedily felt, when the stomach is kept up, by alcoholic stimulants, to a state of extraordinary activity of supply, — and although, in like manner, the habitual use of alcoholic stimulants may cause the stomach to be less susceptible of the loss of the accustomed energy,—yet, when the crisis does come, each condition aggravates the other; the effects of undue disintegration of the nervous matter being more difficult to repair, when the nutritive apparatus is depressed in functional power; and the restoration of the tone of the stomach being impeded by the deficiency of nervous energy, when this has been lowered by excessive action of the brain. The length of time then required for the cure, is proportional to the duration of the causes which have induced the malady; and tedious and difficult is the process of restoration, as every medical man well knows. We shall hereafter have occasion

(§§ 227, 228,) to consider the best methods of medical and hygienic
treatment for this condition; and shall show that the measures
which experience now proves to be the most efficacious means of
restoring the vigour of the system, are precisely such as the physiologist would recommend, under the guidance of the preceding
views of the causation of the morbid state in question.

165. We have thus been led to consider the remote influences of
the prolonged and habitual use of fermented liquors, in however
"moderate" a quantity, upon the digestive apparatus, the excretory
organs, and the nervous system; and we have found that we may
with the highest probability, if not with absolute certainty, attribute
many of the chronic disorders which affect these organs in advancing
life,—especially that loss of functional power, which is frequently
the earliest stage of such disorders, and which, if appropriately
treated in the first instance, might not proceed further,—to the
excessive action to which they have been subjected, under the
stimulating influence of alcoholic beverages.   In so far, therefore,
as the use of these beverages causes or favours such excessive action,
it must in the end be hurtful, rather than beneficial, to the general
health; notwithstanding that its temporary effect may appear to be
wholesome and exhilarating, or at any rate, if negative for good, to
be also negative for evil.   But we have further to consider, whether
·his inference is borne out by the effects of alcoholic liquors taken
habitually in small quantities, upon the functions of circulation and
nutrition.

166. *Effect upon the Circulation.*—It may be difficult to prove
that the ingestion of a *small* quantity of Alcoholic liquor, taken in
conjunction with food, has any decidedly stimulating influence upon
the general circulation; since a certain acceleration of the pulse,
and an increase in its fulness, normally occur during digestion; and
the augmentation produced by the alcohol may be so trifling as to
be scarcely detectible.   Such augmentation, however, is certainly
produced by the imbibition of a quantity usually accounted "moderate;" and we have now to inquire, whether it can recur habitually
through a long series of years, without producing injurious results.
There cannot be a doubt that, in a healthy person, the rate of the
circulation is proportioned to the amount of functional activity of
the principal organs of the body.   We find that it depends in great
degree upon muscular exertion, as put forth in the maintenance of
the erect posture, and still more in active exercise; but it may be
accelerated also by exalted activity of the nervous system, which
sets up an unusual demand for blood in the brain; and its increase
of rate during the digestive process appears to be connected with the

---

¹ *Hygienic*—a proper regulation of exercise, mental and bodily; food,
drink, clothing, bathing, and exposure to pure fresh air.

large supply of blood then transmitted to the chylopoietic viscera,[1] and required for the due performance of their several offices. Now whenever the circulation undergoes any considerable acceleration, there is a tendency to a recurrence of local congestions, arising from the want of power, on the part of the vessels of some particular organ, to allow their current to pass at the same rate with the rest. Of this we have a familiar example in that accumulation of blood in the pulmonary arteries, which is liable to take place in most persons during violent muscular exertion, producing the feeling of being "out of breath;" and which is particularly marked in those, in whom there exists some disordered condition of the lungs, that obstructs the passage of blood through their capillaries.[2]

167. There are few persons, however, in whom there is not some tendency to an *irregularity* of the circulation, which manifests itself in a torpor in some parts, and an undue activity in others. One of the most common forms of this, especially among individuals who work their brains more than their muscles, is a torpor of the current in the extremities, and an undue activity in the cephalic circulation; so that the head is habitually heated, whilst the hands and feet are cold. Now where such is the case, we find that even the normal acceleration produced by the ingestion of food aggravates this disordered condition; so that the face becomes more flushed, and the head more hot after meals, than at any other time.[3] Precisely the same result is observable in such persons, after the use of even a small quantity of alcoholic stimulant; and the habitual production of it cannot be but injurious, as tending to establish that inequality which it should be our endeavour to counteract.

168. Similar inequalities exist in different individuals, in regard to other organs; thus it very frequently happens that the Liver is the part in which a disposition to torpidity of circulation exists; and congestion of its portal system of vessels must stagnate the whole of the circulation through the chylopoietic viscera, from which the blood of that system is derived. Any such disposition to local congestion must operate with increased force in producing general irregularity of the circulation, when the rate of movement is unduly accelerated; just as the outlets to a theatre, which suffice to discharge the entire audience in a few minutes, when the pressure towards them is uniform and regular, are speedily blocked up and produce a stagnation of the entire current, whilst, under the influence of an alarm of fire, every one is rushing toward them with

---

[1] *Chylopoietic viscera* — organs which form chyle.
[2] *Capillaries* — very minute, hair-like vessels.
[3] The acceleration produced by muscular exercise, will of course be un attended by this result; the cause of the acceleration being such as to divert the current from the brain to the limbs, and to make it pass through them with energy and rapidity.

12

undue haste. And as we have seen that hepatic and abdominal congestions are among the ordinary results of excess in the use of alcoholic liquors (§ 155), it cannot be doubted, but that even their moderate employment must aggravate any tendency to such derangement of the circulation, when it already exists. No such derangement can be habitual, and be thus continually liable to aggravation, without laying a foundation for other more serious disorders.—So, again, as we have seen that habitual excess in alcoholic liquors has a tendency to produce determination of the blood towards the Kidneys, and thereby to favour the development of many serious diseases in those organs (§§ 54–58), we can scarcely refuse to admit that where the least tendency to disordered action already exists in them, it must be aggravated by the habitual recurrence of such a slight increase in the afflux of blood to them, as would of itself attract no attention.

169. ·If it be said, that in thus reasoning upon probabilities, we are going further than experience warrants us in doing, we must again take leave to refer to the argument from analogy on which we have already dwelt (§ 156), as a justification of our somewhat theoretical propositions. The whole tendency of modern pathological [1] research has been to show, that the human frame, if endowed with an ordinary amount of inherent vigour, is no otherwise incident to disease, than as it is in various ways subjected to the agency of causes which produce a departure from the normal play of its functions; and that although old age and decay are inevitable, diseases are not, being preventible in the precise proportion in which we are able to discover and eradicate their causes. And when we can clearly trace a relation of cause and effect, between obvious and flagrant violations of the rules of health, and the occurrence of certain forms of *acute* disease, we seem justified in assuming that minor but habitual violations of the same kind must be allowed to participate, at any rate, in the production of *chronic* diseases of the same order. The very nature of chronic disease implies a prolonged action of the causes in which it arises; for no such determinate alteration of the normal functions as it involves, can be at all accounted for by any temporary causes of perversion; —these either inducing a transitory disorder, or, if acting with sufficient intensity, exciting an attack of acute disease. In chronic diseases we find that the organ has, so to speak, *grown to* its perverted action; so that no curative measure is permanently beneficial, which does not first act by withdrawing the cause of the original departure from the healthy state, and by placing the organ in the best condition for its recovery. — We are fully justified, therefore, by all that we know of the causes of disease, in asserting that the

---

[1] *Pathological* — relating to the nature, causes, character and effects of disease.

haoitual use of Alcoholic liquors by healthy individuals, even iu small quantities, is likely, when sufficiently protracted, to favour the development of such chronic disorders; as originally depend upon an irregularity in the movement of the circulating current, or are liable to be augmented by it.

170. *Effect upon Nutrition.* — There appears, moreover, to be an adequate amount of evidence, that the practice in question has an unfavourable influence upon the Nutritive operations, by which the alimentary materials first converted into blood are applied to the regeneration of the living tissues. This influence is not so clearly manifested in the ordinary course of these operations, — which indeed is not demonstrably affected by it, — as in the extra-ordinary demand which is made upon the regenerative powers, for the repair of injuries occasioned by accident or disease. It is well known to Surgeons, that the most desirable of all modes by which the reparation of wounds can be effected, is the simple adhesive process, known as "union by the first intention;" and that where, in consequence of loss of substance, union by the first intention cannot be effected, the most favourable method is that which is termed the "scabbing process;" in which a hard crust being formed upon the surface, so as to protect it from the irritating action of the atmosphere, a continued growth or re-formation of tissue takes place beneath, without any interruption from inflammatory action, until complete filling-up has been effected, and a new cutaneous surface is formed beneath the scab. But it too frequently happens that the reparative processes cannot be induced to take place after either of these fashions, but that inflammatory action is set up in the wound, and matter forms between its lips, or beneath the scab, rendering its detachment necessary, and thus re-converting the wound into an open sore. The healing of this sore must be accom-plished by the much less healthy process of suppurating granula-tion; during the progress of which a large amount of nutritive material runs to waste as purulent discharge, whilst a great degree of constitutional irritation is often set up; and the best termination of which is the formation of a cicatrix, that subsequently undergoes an unsightly and often inconvenient contraction, from which the new tissue formed under a scab is free.

171. Now the occurrence of the first of these modifications of the healing process, is an obvious indication of such a healthful condition of the nutritive operations, as can repair the effects of an injury in the most complete manner, with the least possible waste of nutritive material, and with the most entire absence of constitu-tional disturbance. Whilst, on the other hand, the impossibility of procuring it, even under the most favourable circumstances of rest, fresh air, and wholesome aliment, indicates that the nutritive functions are *not* in their normal condition. Amongst the lower

animals we seldom find injuries repaired in any less favourable mode, unless the part be placed in circumstances adverse to this healthy action. But among " civilized" communities of men, the case is very different; for the occurrence of the scabbing process, in the case of any but trivial wounds, is the exception, not the rule, — being, in fact, so rare that many Surgeons never think of attempting to bring it about. Now that there is nothing essentially different in the constitution of Man, which places him in this respect at a disadvantage as compared with the lower animals, appears from the fact that all who have visited "savage" nations, in whom more constant exposure to air is practised, and who enjoy immunity from many causes of disease which exist in civilized communities, have been struck with the facility with which wounds heal among them, and with their remarkable freedom from that constitutional disturbance, which amongst ourselves almost invariably follows severe injuries. Thus Hawkesworth in his voyage to New Zealand makes particular mention of " the facility with which wounds healed that had left scars behind them, and that we saw in a recent state; when we saw the man who had been shot with the musket ball through the fleshy part of the arm, his wound seemed to be so well digested, and in so fair a way of being perfectly healed, that if I had not known no application had been made to it, I should certainly have enquired with a very interested curiosity after the vulnerary herbs and surgical art of the country." Of these people, he states that at that period water was their sole and universal liquor.

172. Now, it would be absurd to maintain that the habitual moderate use of fermented liquors is the *sole* reason of the rarity of this healthful operation of the reparative process amongst ourselves; since a multitude of other departures from the laws of health are continually practised by almost every member of a civilized community. But if we look to the unquestionable fact, that habitual excess in the use of fermented liquors produces a condition altogether *opposed* to the healthful performance of these processes, so that the slightest scratch or abrasion may give rise to a rapidly-fatal attack of inflammation (§ 63), it can scarcely be denied that where a minor departure from the normal condition shows itself, and the same cause has been in action in less intensity, that departure may be reasonably considered, in part at least, as its effect. And this conclusion is remarkably confirmed by the surgical experience of the late campaigns in India, on occasions on which there had been, from accidental causes, an interruption in the usual supply of spirits. Thus Mr. Havelock, in his " Narrative " in reference to the wounded, after the victories in India, observes :—" The Medical Officers of this army have distinctly attributed to their previous abstinence from strong drink the rapid recovery of the wounded at Ghuznee." And

Mr Atkinson, in his work on Affghanistan, is more explicit, stating that "all the sword cuts, which were very numerous, and many of them very deep, united in the most satisfactory manner; which we decidedly attributed to the men having been without rum for the previous six weeks. In consequence, there was no inflammatory action to produce fever and interrupt the adhesion of the parts."

173. From the foregoing considerations, then, we seem entitled to draw the general conclusion that, in the "average man," the habitual use of Alcoholic liquors, in moderate or even in small quantities, is not merely unnecessary for the maintenance of bodily and mental vigour, but is even unfavourable to the permanent enjoyment of health, even though it may for a time appear to contribute to it. For, as it is justly remarked by Dr. Robertson, "that man only is in good health, who recovers rapidly from the simple accidents incidental to his occupation, and from the simple disorders incidental to his humanity and to the climate he lives in, and who can bear the treatment that those accidents or those disorders demand;" and if such be not the case, we may feel confident, that however great the temporary power of exertion may be, such power is destined to give way at a period much earlier than that of its normal duration. And if it be true, as we have endeavoured to show, that the effect of the habit is not merely to induce certain predispositions to disease by its own agency, but also to favour almost any of those which may already exist in a latent form, we have an additional right to affirm, that even the most moderate habitual use of alcoholic liquors becomes to the "average man" positively injurious, if protracted for a sufficient length of time to allow of the development of its effects.

12 *

CHAPTER III.

ARE THERE ANY SPECIAL MODIFICATIONS OF THE BODILY
OR MENTAL CONDITION OF MAN, SHORT OF ACTUAL DISEASE,
IN WHICH THE OCCASIONAL OR HABITUAL USE OF ALCOHOLIC
LIQUORS MAY BE NECESSARY OR BENEFICIAL?

174. There appear to be three classes of cases, in which recourse
may be had with temporary advantage to the use of Alcoholic
liquors; those, in the *first* place, in which there is a demand for
some extraordinary exertion of the animal powers, and in which the
occurrence of subsequent depression may not be an adequate objection to the employment of a stimulus that enables the system to
meet it; those, in the *second* place, in which there is a deficiency of
the proper sustenance, and in which alcohol serves as a heat-producing article of food; and those, in the *third* place, in which there
is a want of sufficient vigour on the part of the system itself, to digest
and assimilate the aliment which it really needs for its support.

I. — DEMAND FOR EXTRAORDINARY EXERTION.

175. Of the *first* class, the following appropriate example may be
extracted from the letter of Dr. J. D. Hooker, already cited, "I
know of only one occasion," he says, "on which the use of spirits
appeared indispensable; and that was, when a little more exertion
at the crowning of a mighty and long-continued effort was demanded.
Thus the ship, when sailing in the pack-ice, is sometimes beset, or
falls to leeward into the lee-ice. This takes two or three minutes;
but if there is much wind, it takes many hours to get her out. Not
being in command, the sails are of no use; and the ice prevents her
from moving in any way but with it to leeward. Under these circumstances, the only way to get her out, is by fastening ropes from
the ship to the larger masses of ice, and warping her out by main
force against the wind. Now I have seen every officer and man in
the ship straining at the capstan for hours together, through snow
and sleet, with the perspiration running down our faces and bodies
like water. Towards the end of such a struggle, at the mighty
crowning effort, I have seen a little grog work wonders. I could not
have drunk hot coffee without stopping to cool; nor if I had, do I
think it would have supplied the temporary amount of strength
which was called for *on the spot* under circumstances like this.
These, however, are extreme cases, which do not affect the sailor
in his ordinary condition, and which any ship might be well prepared for."

176. It must be within the experience of most persons, that a very small quantity of Alcoholic stimulus has been of similar efficacy in sustaining the nervo-muscular energy under some tempo-rary effort, which circumstances called for, and to which the system, exhausted by previous fatigue, would not otherwise have been equal. And the writer can speak from his own knowledge of its corres-ponding effect, in quickening and freshening the mental power, during a brief period through which it could not otherwise have been sustained. Of course, in every such case, a corresponding de-pression is subsequently felt ; but this depression is rather traceable to the fatigue of over-exertion, than to the re-action consequent upon over-excitement. For, in the cases alluded to, the effect of the alcoholic liquor is not to quicken the circulation, or to exalt any of the functions above their normal activity, but merely to keep them up to par : and its use for such a purpose is therefore free from many of the objections, which have been urged against its habitual employment.

177. But it must not hence be supposed that recourse to Alco-holic liquors can habitually be had with impunity for purposes of this kind. Every kind of "forcing" must be in the end injurious to the vital powers, and more especially to those of the nervous system ; and the more frequently and violently it is practised, the more speedily may we expect that functional derangement will manifest itself. Extreme over-tasking of its powers is often so im-mediately followed by apoplexy, paralysis, epilepsy, mental de-rangement, or fatuity, that no one has any hesitation in regarding these as the natural results of the previous immoderate exertion ; and we appear equally justified in attributing similar results to similar causes, however remote the results may be, where causes less potent have been in continual or frequently-repeated operation. For every such irregularity tends to derange the nutrition of the system ; and if a renewal of the irregularity should take place before the effects of the preceding derangement have been recovered from, they are of course aggravated ; and thus a cumulative result is produced, and a permanently-disordered state of nutrition estab-lished, which manifests itself at last in some serious and settled form of cerebral disease.

178. The case resembles that of the racer, excited to put forth his utmost speed, or the jaded roadster goaded to a temporary im-provement of his pace, by the application of the spur. The spur gives no strength ; but, like the dram to the sailor toiling at the capstan, or the glass of wine to the public speaker wearied with his previous exertions, it calls forth the most vigorous exercise of the remaining strength The racer may fall dead on the spot ; the roadster may sink from exhaustion ; but the spur has only been the indirect means of bringing about this catastrophe, the real cause of

it being the undue exertion which it has called forth. And in like manner, when recourse has been had to alcoholic liquors, for the maintenance of the power to meet some extraordinary demand upon the bodily or mental energy, and the amount used has been merely such as to meet that demand, we ought to attribute the subsequent exhaustion rather to the violence of the effort which has been put forth, than to the stimulus, trifling in itself, by which the system was rendered capable of making it. The occasional dram or glass of wine would of itself have produced but little mischief in comparison; and its consequences might have been manifested in some other way. But the frequent over-exertion of the vital powers, especially those of the nervous system, must ultimately *tell* upon the fabric, under whatever kind of excitement it is called forth.

179. However desirable, then, it may be to avoid the necessity for such immoderate exertion, it can scarcely be denied that occasions will arise in the experience of some persons, in which the temporary assistance derived from Alcoholic liquors could scarcely be replaced by any other. When the choice lies between the easy and satisfactory performance of the prescribed duty, and the discharge of it as a task which *must* be got through at all hazards by the most determined bracing-up of the powers for its execution, there can scarcely be a doubt, in the opinion of the writer, that if the former can be procured by the use of such a small dose of alcohol as shall merely raise the vital powers for a time to their usual energy, it will be followed by *less* of subsequent exhaustion than the latter. But again, he would repeat,—and he cannot do so too often, or too earnestly,—that the *habitual* recourse to such a practice is fraught with the greatest prospective danger; since it encourages the delusive idea, that the exertion which is thus for a time sustained, is really doing no injury to the system; besides which, it is next to impossible that the frequent use of alcoholic liquors, however moderate, can be persevered in, for any length of time, without favouring the production of that disordered state of nutrition of the brain, which the irregular activity of the nervous system has of itself so marked a tendency to generate. It should rather be the aim of those who have accustomed themselves to such assistance, to avoid the necessity (so far as may be possible) for such extra-exertion; and to prepare themselves to meet it, when it is indispensable, by careful and constant attention to all the rules of health. The most beneficial results from such a use of stimulants, are to be experienced by those who are habitually abstinent; since the quantity of alcoholic liquor which they require for the purpose is extremely small; and whatever injurious effects it may produce will be more likely to be dissipated, when a considerable interval elapses before it is again resorted to. When alcoholic liquor is employed as an ordinary beverage, the quantity required to give the

desired aid, on the occasions in question, is such as must of itself
exert a prejudicial influence on the system.

180. Nearly allied to the preceding cases, are those in which the
use of Alcoholic liquors may be found beneficial, in assisting to fortify the system against a *temporary* exposure to cold or damp, separately, or in combination. We have already examined into the
reputed efficacy of alcoholic liquors in favouring the resistance to
cold; and have found reason to adopt the conclusion, that this reputation is altogether fallacious as regards the power of *continued*
endurance. There can scarcely be a question, however, that although, considered simply as a heat-producing material, alcohol is
inferior in some important particulars to such oleaginous matters as
can be readily introduced into the current of blood, it has *for a time*
the power of keeping off the chilling influence of severe external
cold, in virtue of the augmented rapidity of the circulation which it
induces, and particularly of the determination of blood which it
favours towards the vessels of the skin. And this effect seems to
be exerted with still greater benefit, when cold and damp are acting
together; their depressing influence being kept at bay for a time by
the moderate use of alcoholic stimulants, so that no injurious result
is subsequently felt from an exposure which might otherwise have
been followed by a severe "cold," an attack of rheumatism, or some
other malady, as determined by the idiosyncracy[1] of the individual.

181. It is not here argued, however, that Alcoholic liquors afford
the *best* means of resisting such influences. On the contrary, it is
within the experience of most persons, that muscular exertion, where
it can be employed, is a far better means of keeping up that vigour
of the circulation which shall resist the influence of the external
chill, than the use of any stimulants whatever in a state of bodily
inactivity. But where circumstances prevent a resort to the former,
and the choice lies among the best internal means of protection,—
as in the case of a traveller exposed to cold and wet on the top of a
coach,—we seem justified in believing that if the chilling influence
is powerful and likely to be of short duration, it may be better resisted by a stimulating dose of alcoholic liquor, than in any other
way. But if the resisting power is to be prolonged, such a course
is most undesirable; for the system is never so obnoxious to the
depressing influence of cold and damp, as when it is already in a
state of depression resulting from previous over-stimulation; and
the use of coffee, cocoa, and other hot beverages, with solid food,
which shall aid in permanently sustaining the heat of the system, is
then unquestionably to be preferred. Here, again, we would remark,
that the habitual abstainer has decidedly the advantage, since a very
small amount of the stimulus is sufficient, as in the former case, to

---

[1] *Idiosyncracy* — constitutional peculiarity.

produce the desired result; and that if recourse be too frequently had to it, the remote consequences of alcoholic excitement may be expected to manifest themselves.

II.—DEFICIENCY OF OTHER ADEQUATE SUSTENANCE.

182. The *second* class of cases, in which the use of a small amount of alcoholic liquors seems beneficial, or at any rate justifiable, is that in which there is a deficiency of the proper sustenance, so that the alcohol supplies the means of maintaining the animal heat, for which the animal tissues would otherwise be attacked. Under such circumstances, too, the temporary elevation of the habitually-depressed state of the animal power seems rather beneficial than injurious. Of this we have a remarkable example in the well-known case of the Mutiny of the *Bounty;* from Captain Bligh's Narrative of which the following passages are extracted. "At daybreak I served to every person a tea-spoonful of rum, our limbs being so much cramped that we could scarcely move them." Further on— " being unusually wet and cold, I served to the people a tea-spoonful of rum each, to enable them to bear with their distressed situation." And again,—" our situation was miserable ; always wet, and suffering extreme cold in the night, without the least shelter from the weather. The little rum we had was of the greatest service ; when our nights were particularly distressing, I generally served a tea-spoonful or two to each person, and it was always joyful tidings when they heard of my intention." Now, however decidedly we may give the preference to hot tea, coffee, or cocoa, with plenty of nourishing food, over alcoholic liquors, in facilitating the endurance of such an exposure, it can scarcely be questioned that in circumstances such as those of Captain Bligh's crew, the administration of the few drops of spirit was of the most important service, both as supplying combustible material, and as enabling the powers of their system, already seriously depressed, from being fatally reduced by the privations to which the party was subjected.

183. The beneficial influence of a small quantity of Alcoholic stimulus, in contributing to the endurance of bodily labour under circumstances peculiarly trying, and under the disadvantage of a deficient allowance of animal food, has been demonstrated on an extensive scale by the hygienic [1] experience of the large prison at Nismes, called the "Maison Centrale," of which an account has been recently published by the chief physician, M. Boileau Castelnau, who has been connected with the prison for the last twenty-five years.[2] Of this account an abridgment will be here given, as the

---

[1] *Hygienic* — relating to the causes by which the health of the system is maintained and improved

[2] See the Annale. d'Hygiène Publique, Jan. 1849.

facts are considered by the writer as of very great importance, in disproving, by the experience of a large number of individuals, the position of those who assert that under no circumstances can the habitual use of alcoholic liquors be otherwise than injurious. This prison usually contains a population of 1200 convicts, most of them adults, the minimum age being eleven. Its wards have been habitually over-crowded and ill-ventilated, and insufficiently heated in winter; and the food of the prisoners has been coarse and innutritious, whilst more labour has been exacted from them than their strength has been adequate to perform. The prisoners, moreover, have been subjected to the tyranny of brutal keepers, frequently loaded with irons, and occasionally severely whipped. Under these circumstances, it is not surprising that the rate of mortality in the prison has been always high, varying from 1 in 23·88 to 1 in 7·85, whilst the average rate of mortality amongst the inhabitants of the town of Nismes, of the same age and sex, amounted to 1 in 49·9. The rate of mortality in the prison underwent considerable variations in different years; and for these variations some definite cause could generally be assigned. Thus the winters of 1828–9, and of 1829–30 were unusually severe and prolonged; and the rate of mortality for 1829 was 1 in 9·40, whilst for 1830 it was 1 in 8·50; clearly proving the fatal influence of a low temperature upon systems debilitated by insufficient food, impure air, and work disproportioned to their strength. With the exception of the year 1833, in which the mortality was again great, the rate was much less for several subsequent years, varying between 1 in 11·35, and 1 in 15·62; but in 1839 it suddenly rose from 1 in 12·32 to 1 in 7·85. The cause of this terrible augmentation (from 102 deaths to 162) seems to have lain in a ministerial ordinance issued on the 10th of May, 1839, limiting the alimentary articles allowed to be sold at the canteen to potatoes, cheese, and butter. Previously to that time, the convicts had it in their power to lay out a portion of their earnings, which was at their own disposal, in the purchase of wine and tobacco, in addition to the articles just named; but these were now prohibited.

184. "In order fully to appreciate," says Dr. J. Coxe (by whom this case is cited in the ninth edition of Dr. A. Combe's Physiology of Digestion), "the amount of misery thus entailed upon the prisoners, the reader must be aware that in the south of France, wine is considered an absolute necessary of life. It is drunk by the poorest of the people, and appears essential to enable them to digest their coarse unstimulating food. Within the town of Nismes, it costs about a penny the litre, (1¾ pint); and without the wall, where it is free from duty, the labourer may drink it at a penny the hour. Potatoes, butter, and cheese, could not replace its stimulus; and besides, the south of France containing no pastures, the

butter was bad and dear; and the cheese also dear. Hence the pittance at the disposal of the prisoners was more than ever insufficient to supply the deficiency of nutriment."

185. The rate of mortality was somewhat diminished in the following year, the diet being increased, and the prisoners receiving small supplies of wine and tobacco at exorbitant rates, principally through the connivance of the officials; still, however, it remained very high, the number of deaths in 1840, being 135 out of 1216, prisoners, or 1 in 9·07. Attention being now attracted to the condition of the prisoners, an attempt was made to ameliorate it; the old keepers being removed, and their places being supplied by the " Frères des écoles chrétiennes," [1] who substituted moral persuasion for physical force; and the diet being improved both in quantity and quality. In consequence of these measures, the mortality again began to diminish, and reached its lowest point in 1844, when the number of deaths was only 56 out of 1290 prisoners, or 1 in 23·88. The " Frères," however, being disgusted at the continual obstructions which their measures received, gave up their charge; the old system of hard work and cruel punishments was again introduced; and the pittance at the disposal of the prisoners was diminished to a mere fraction. The effect of this change speedily showed itself in the increased mortality, the average of deaths progressively increasing in the years 1845, 1846 and 1847, to 1 in 19·63, 1 in 16·52, and 1 in 13·57. One of the first acts, however, of the revolutionary government of February, 1848, was to put a stop to the system of convict-labour, as it was then carried on, and the result of this change was speedily apparent in the diminished mortality; for whilst the number of deaths during the seven months ending October 31st, 1847, had been 44, only 16 deaths took place during the corresponding months of 1848.

186. Now the principal lesson taught by this fearful history, is the dependence of the vital powers upon food, and the fatal effects of the exaction of severe labour from men insufficiently supplied with aliment, especially when they are subjected to the additionally injurious influences of a low temperature, foul air, and ill treatment. But it seems obvious from the large increase in the rate of mortality which ensued upon the prohibition of *wine* — no extraordinary depression of temperature having existed to account for it, — that its deprivation exerted a positively injurious effect. If an adequate measure of nutritious food had been supplied in its stead, the change would doubtless have been for the better; but the support given by the wine, which was probably too weak and poor to have any decided stimulating effect in moderate quantities, had be-

---

[1] *Frères des écoles chrétiennes* — brothers of the Christian schools — a religious order.

come so necessary to the debilitated systems of these men, that its withdrawal was fatal to many among them.

187. From these two cases, then, and from others which might be cited to the same effect, we seem justified in concluding, that the use of Alcoholic liquors in small quantity may assist in sustaining the powers of the system, when these have undergone an extreme depression from the combined influence of exposure or exertion, and of want of food; so that under such circumstances the alcohol does decidedly more good than harm, and should therefore be employed when accessible. And this we may freely admit, without having in the least degree to qualify the doctrine previously advanced, that continual exposure and protracted exertion may be better sustained without the use of alcoholic liquors than with it, when an adequate supply of wholesome food is to be had, and the stomach is capable of digesting it.

III. — DEFICIENCY OF CONSTITUTIONAL VIGOUR.

188. We have now to inquire into the *third* class of cases, in which a temporarily beneficial result appears to be derived from the occasional, or even (for a time at least) the habitual use of Alcoholic liquors; — those, namely, in which there is a want of sufficient vigour on the part of the system itself, to digest and assimilate the aliment which it really needs. Such cases present themselves in all ranks of life. In the higher, they too frequently result from heated rooms and late hours, from the want of regular exercise of mind and body, and from habits of self-indulgence and "coddling," which foster, especially in females, what may have been an hereditary weakness of digestive power. In the middle classes, it is usually traceable to the "wear and tear" of professional or commercial avocations; to undue cerebral labour, carried on, as this frequently is, in ill-ventilated apartments; and to the anxieties incident to the conscientious discharge of the duties of a profession or to the fluctuations of business. Among the lower classes, on the other hand, it is traceable rather to the condition of their dwellings, workshops, and persons; to the want of ventilation of the buildings in which they dwell or labour; to the miasmatic atmosphere of their ill-drained streets, and to the foulness of their skins and garments.

189. Now in the first of these groups, it is obvious that the want of appetite is a natural result of the reduction of demand for aliment to its lowest point; for where neither the muscular nor the nervous systems are adequately exercised, and where the body is habitually kept in a temperature not far below its own, there can be very little "waste" to be repaired, and a very small amount of combustive action can be needed to keep up the heat of the body to its proper standard. But the digestive powers are very liable,

13

when their natural use is too little called for, to sink *below* the level at which the demands of the system should keep them; and thus an almost total want of appetite, and extreme debility of the stomach, are the result, which of course tends to augment the habits of self-indulgence, and to foster the whole system of "coddling." In such cases, an apparent benefit is derived from the habitual employment of a glass or two of wine or a tumbler of bitter ale; but this merely facilitates the persistence in a wrong course; and every judicious practitioner would now assent to the truthfulness of the advice given by Abernethy in a case of this kind, to "live on a shilling a day and earn it." It is utterly impossible that alcoholic liquors can counteract the influence of heated rooms and late hours; that they can stand in the place of healthful exercise of mind and body; or that they can neutralize the evil results which are sure to proceed from the habitual direction of the attention to self. All that they can beneficially do is, to create for a time that appetite which ought to be naturally felt, and to urge the unwilling stomach to digest that food which the body really requires. But this they can only effect by their *stimulating* properties; and as the usual dose almost invariably ceases after a time to exert its original influence, it requires a gradual increase, until the evil effects of its habitual use in such a state of the system are unmistakeably manifested.

190. The true cure for conditions of this kind lies in such an entire change of habits, as shall place the system in the condition most favourable to the recovery of its vigour, or to the acquirement of that which it has never enjoyed;—the substitution of fresh air and bracing breezes, for heated and ill-ventilated rooms; of early and regular hours, for the system of turning night into day and day into night; of plain but wholesome fare, for seasoned dishes and refined cookery; of the use of even a weakly pair of limbs, for that of a carriage and horses; and of labour in behalf of others, for the weariness of ennui or continual thought of one's-self.

191. Nevertheless it may happen that after all these means have had a fair trial, and considerable improvement may have been produced, the stomach may not be equal to its work; and this is liable to be the case more particularly with those, to whom weak digestive powers have been transmitted from their parents (generally in consequence of their own unhealthful habits), or in whom they have been fixed (so to speak) by an erroneous system of bodily and mental training, and especially by the habitual use of stimulants during childhood and youth. In such cases the writer believes that the habitual use of a small quantity of alcoholic stimulant, especially when combined with a bitter tonic, may be of more service than any other form of medicine; and if care be taken not to employ it to such an extent as to produce an artificial appetite, or to

force the stomach to digest more than the system really needs, it does not appear likely to have the same permanently injurious effects, as it exerts in most other cases. It will generally be found to be an indication of its beneficial use, that the dose does not require increase; the small quantity originally taken, continuing to exert its good effects; and this benefit will be more likely to be persistent, if the use of the alcoholic stimulant be intermitted, whenever the digestive powers seem adequate to the support of the system without it.

192. The want of appetite and feebleness of digestive power, so common among individuals in the middle classes, who go through an undue amount of cerebral labour, frequently under circumstances which are of themselves prejudicial to health, has been already adverted to under another head (§§ 163, 164); and it has been shown that the use of alcoholic liquors cannot in general be regarded as likely to be permanently beneficial in such a condition, although temporary benefit may doubtless be derived from it. It is impossible that Alcohol can supply the place of mental repose to the man whose intellect is over-tasked, and whose anxieties are unduly excited for himself or for others; or that it can be an efficient substitute for muscular exercise to the man of sedentary habits, or for fresh air to him who is habitually exercising his brain in a close ill-ventilated apartment. All that it *can* do, is, as in the former case, to restore the appetite which ought to be felt, and to force the digestive powers to the discharge of the duty which they are indisposed to perform of their own accord. And here, too, we find that when stimulants are habitually employed for such a purpose, they gradually lose their power; and the wearied stomach, like the jaded roadster, at last breaks down, under the combined influence of the withdrawal of nervous agency consequent upon cerebral exhaustion, and of the depression of its own energies consequent upon the habitual over-excitement to which it has itself been subjected.

193. Here then, it is obvious, that the use of Alcoholic stimulants can only serve as a palliative; and that the true remedy can only be found in such change of habits, as shall bring back the system as nearly as possible to the natural state. The intellectual labour must be moderated; the mind must be prevented from dwelling on its own sources of anxiety, by the healthful influences of social and domestic intercourse, of variety of occupation, and of objects that shall interest without exciting it; and the body must be placed by regular exercise, fresh air, and adequate repose, in the most favourable condition for the endurance of mental labour. Such measures steadily pursued, with an occasional complete intermission from the ordinary occupations, and an entire change of scene with the accompaniment of fresh objects of interest (for *ennui* is to be especially avoided), more especially when a bracing air and aug-

mented muscular exercise tend still further to the bodily invigoration, will usually be found sufficient, when employed in time, for sustaining the appetite and digestive powers under that amount of mental labour to which the system is really equal; and recourse should be had to all such natural means of procuring and sustaining the vigour of health, before the artificial and delusive aid of alcoholic stimulants is invoked. It is, indeed, among the most injurious results of their habitual use, that it is found possible through their means to prolong the health-destroying system; and thereby, like the trader who bolsters up his failing credit with accommodation-bills, to carry onwards, from page to page of the book of life, a heavy balance which *must* be accounted for at some subsequent period.

194. Still the writer is by no means disposed to deny, that after all other practicable means have been taken for the invigoration of the system, the habitual use of a small or moderate quantity of Alcoholic liquors may be found beneficial in some individuals of the class referred to; enabling them to digest that food which the system really needs, and thus contributing to sustain their powers under an amount of exertion, to which they would not otherwise be equal. And this will be especially the case (as with the class first treated of), where, from hereditary predisposition, or the habits of early life, there is a fixed constitutional debility of the digestive powers. In such instances, the stimulating effects of the alcohol do not manifest themselves; it is not found requisite to increase the dose; and the practice is continued with apparent benefit through the whole of life. A characteristic example of the results of experience in this respect is afforded by the case of the late Dr. Joseph Clarke of Dublin, who lived to the age of 76 years, and who discharged the duties of a laborious profession, with scarcely an intermission to the end of his life.[1]

195. The craving which is felt for Alcoholic liquors among the classes whose labour is rather physical than mental, and the benefit which in many cases appears to be derived from it, proceeds from a different cause. Nothing can be conceived in itself more likely to whet the appetite and invigorate the digestive powers, than regular but not excessive muscular toil, with that moderate occupation of mind which the execution of the labour involves; but in order that this may exert its proper effect, it must be carried on under circumstances otherwise favourable to health, and more especially in a pure atmosphere of moderate temperature. If, in place of this, the air be already loaded with carbonic acid, an obstruction is created to the unusually rapid exhalation of that gas which muscular exertion

---

[1] See the sketch of Dr. Clarke's Life and Writings, by his Nephew, Dr. Collins, p. 81.

involves; and the labour cannot be borne without the assistance of stimulants. And if the atmosphere of the dwelling be charged with the noxious emanations resulting from animal or vegetable putrefaction, the appetite and the digestive powers fail, the aliment which the system really needs for the regeneration of its "wasted" material is no longer prepared and supplied to the circulating current, and the strength consequently flags. Under such circumstances, recourse is had with apparent benefit to the use of alcoholic liquors; for they spur on the stomach to its work, and cause it, for a time at least, to furnish what is needed for the maintenance of the various functions of the body; the whole train of which depends, more or less directly, upon the due performance of the digestive operation.

196. Now, it is a remarkable characteristic of this condition, that the stimulus which was at first found sufficient very speedily ceases to produce its usual effect; and that the feeling of necessity for it increases, the more it is used. Of this, we have an example, — which, though an extreme case, teaches the lesson with the force that extreme cases alone can do, — in the condition of the journey-men Tailors employed in the large London workshops, as disclosed by the inquiries whose results are published in the first "Sanitary Report" (1842). The heat and closeness of the workshops were stated by the witnesses to be such, that on the coldest nights of winter, large thick tallow candles melted and fell over with the heat; and fresh hands from the country fainted away. In order to get the strength up for the day's work, and to create an appetite for break-fast, it was customary to take a glass of gin at seven o'clock in the morning; and this was repeated three or four times in the subse-quent ten hours. Now the utter inability of the alcoholic stimulus to afford more than a temporary power of endurance under such a state of things, and the cumulative effect of the noxious atmosphere on the one hand, and of the habitual use of spirits on the other, are fearfully shown in the excessive mortality among this class of men, especially from consumption; their average age not being above thirty-two, and a man of fifty being considered as superannuated.[1]

197. Nothing can be more absurd, then, than to maintain that any real benefit is derived from Alcoholic liquors in such cases, or that it can in the least degree supply the place of pure air, or enable the body to resist the influence of excessive heat. Nor can it be more potent in preventing the morbific influence of putrescent miasmata; nor again, can it make up for the want of personal cleanliness. These agencies can only be remedied by their proper antagonistic measures; — hot and foul air by proper ventilation;

---

[1] The writer has been informed that these workshops have been greatly improved of late years, especially in regard to ventilation; and that the craving for spirits, on the part of those employed in them, has gradually ceased to manifest itself.

noxious emanations from the soil by efficient sewerage; filthiness of the skin and garments by the use of baths and wash-houses;—and if they be allowed to continue, they *must* exert their influence on the bodily system, all the alcohol in the world notwithstanding. When, on the other hand, they are removed, — the artizan's labour being prosecuted in pure air, and his home and garments being kept clean and fresh, so that his skin and lungs are allowed their due exercise, — it will be seldom, if ever, that anything else will be required to sharpen his appetite, and invigorate his digestive powers, for the consumption of as much food as his system may require.

198. On the whole, then, we may conclude that in by far the greater number of cases falling under one or other of the above categories, the influence of the habitual use of Alcoholic liquors, while it may seem temporarily beneficial, is in the end rather pernicious than otherwise; and this not so much (in the cases now under consideration) by their own specific effects on the system, as by causing the individual to *feel* less need of the very change which is needed for the restoration of the body to its wonted vigour. The insensibility to the effects of various morbific causes, which the habitual use of these stimulants induces, and the toleration of them which it thus permits, may be regarded, indeed, as one of its most injurious results. Those who are prevented from feeling the *immediate* consequences of their improper course, flatter themselves that they are uninfluenced by them, and give to their wine, their spirits, or their beer, the credit of the escape. But this is far from being the case. The enemy is only baffled, not dispersed; and although he lies concealed for a time, he only waits until his onslaught may be more effectually made. Any systematic departure from the laws of health — all experience teaches — *must* exert its influence on the system, sooner or later; the sooner it does so, the more readily may the mischief usually be corrected; whilst the postponement of its effects tends to render the process of cure as protracted as the operation of the causes has been. It is one of the greatest benefits of the abstinent system, then, that by making the evils of such a departure less endurable, it sooner prompts the sufferer to seek a remedy.

198. *Pregnancy.* — Among the modifications of the bodily condition, short of actual disease, in which the occasional and even the habitual use of fermented liquors *seems* desirable in some instances, are the states of pregnancy and lactation. The state of *pregnancy* frequently occasions a peculiar irritability of the stomach (apparently of a purely nervous character), which indisposes it to retain the nutriment really required by the system, or which prevents it from properly digesting and preparing it when retained. This irritability is occasionally so aggravated, as to become the subject of medical

treatment; and the most powerful sedative medicines [1] are sometimes required to subdue it sufficiently for the retention of even
small quantities of food. Sometimes even these are ineffectual; and
more relief is obtainable from small quantities of wine, frequently
repeated, than from any thing else. Dr. Meigs (of Philadelphia)
mentions a case in which nothing could be borne but champagne.
In milder cases of the same kind, it often happens that a small
quantity of fermented liquor, taken with the principal meal,—seems
to establish a *tolerance* of it in the stomach, and to promote its
digestion, in a way which no ordinary sedative or tonic medicine can
effect; and it certainly seems a less evil to employ this, even habitually, during the period of pregnancy, than to allow the system,
both of mother and fœtus, to be suffering for want of the aliment
which this condition so peculiarly requires. And as the source of
irritation is temporary, there is less danger than in other circumstances, lest the demand should be rendered permanent, by the
habituation of the stomach to the stimulus.

199. But the evils attending its habitual use, even under such
circumstances, can only be reduced to their minimum, by very
careful attention to all the other conditions favourable to health
during the pregnant state — especially fresh air, moderate exercise,
early hours, adequate repose, and the avoidance of all sources of
excitement; and also by the strict limitation of the quantity of the
alcoholic liquor to that which is sufficient to produce the desired
result. The writer has known cases, in which, under such watchful
regulation, great benefit appeared to be derived from the very
moderate use of alcoholic liquors, (especially of those in which the
bitter and sedative properties of the hop are combined) without any
corresponding disadvantage; the stomach being thereby enabled,
so long as the pregnant state lasted, to receive and digest the food
which the system really needed; and the requirement not being
felt after its termination. But it must not be forgotten that the
habit of indulgence in fermented liquors, once established, is often
felt by females, as well as by men, to be very difficult of relinquishment; and where there is reason to believe that the individual does
not possess self-command sufficient to break through the habit at
the proper time, it might be advisable to endeavour to substitute a
medicine for a *beverage*, giving to the alcoholic compound such a
form as may render it not peculiarly palatable or inviting.

200. *Lactation.* [2] — The benefit derivable from the use of Alcoholic liquors to support the system during *lactation*, is more doubtful. Certainly it may be affirmed that in every case in which the

---

[1] *Sedative medicines*—medicines which reduce the vital actions.
[2] *Lactation* — the period of suckling; the nourishment of the child from
its mother's breast.

appetite is good and the general system healthy, the habitual use of these stimulants is no more called for, than at any other time; and that they are likely to produce the same injurious effects, as when unnecessarily given under ordinary circumstances. The regular administration of alcohol, with the professed object of supporting the system under the demand occasioned by the flow of milk, is "a mockery, a delusion, and a snare." For alcohol affords no single element of the secretion; and is much more likely to impair than to improve the quality of the milk. The only mode in which it can contribute, even indirectly, to increase the amount of solid aliment which the secretion may contain, is by affording a supply of combustive material, the consumption of_ which may leave more oleaginous and saccharine matter to pass into the milk. But where the appetite already prompts to the *ingestion*, and the stomach is equal to the *digestion*, of an adequate amount of solid food, no such benefit can be looked for; and although it cannot be certainly affirmed that the character of the milk is *always* impaired by the habitual use of moderate quantities of alcoholic liquors, yet there can be little doubt that such is *usually* the case. For it is unquestionable that their *excessive* employment is highly prejudicial to the quality of the milk, and thereby to the health of the child; having a special tendency to occasion derangements of the digestive organs, and convulsive complaints. [2] This, indeed, might be fully expected; since all that we know of the mode in which substances taken into the blood affect the mammary secretion, would lead us to expect that alcohol, if introduced into the circulation more rapidly than it can be consumed, would pass into the milk, and would consequently produce the same effects upon the child as if directly given to it,—besides deranging by its presence, the act of secretion itself, in virtue of its tendency to produce coagulation of albuminous matters. And the fact that multitudes of women of good constitutions, whose general habits are conducive to health, go through the period of lactation without any feeling of debility,—simply finding their appetite increased during its continuance,—is a sufficient proof that this condition is not one, which in itself occasions a demand for alcoholic liquors.

201. But there are cases in which, notwithstanding all that can be done to promote the general health, the stomach does not seem capable of retaining and digesting the requisite amount of nutriment, except under the artificial assistance afforded by Alcoholic liquors; and in which it appears more desirable, for the welfare alike of mother and child, that such assistance should be afforded, than that lactation should be carried on without it. In one case of

---

· Dr. North says (Practical Observations on the Convulsions of Infants) that he has seen these almost instantly removed by the transference of the child to a temperate woman.

this kind that fell particularly under the writer's notice, in which the mother was most anxious to avoid the assistance of fermented liquors, and began to nurse without their support, the milk was obviously too poor in quality, and not sufficient in quantity, for the nutrition of the infant; and the use of a single glass of wine, or a tumbler of porter, per day, was followed by a speedy and marked improvement in the condition of both mother and child; and this small allowance did not require to be increased during the continuance of the lactation, and was relinquished without difficulty soon after the weaning of the infant. In such cases the alcoholic liquor seems to have no other operation, than that of enabling the stomach to digest the amount of solid aliment required by the system; whilst the smallness of the quantity of alcohol introduced at any one time, prevents it from either itself passing into the milk, or exerting any injurious influence on the secreting process. But it may be questioned whether the practice is *in the end* desirable; or whether it is not, like the same practice under other circumstances already adverted to, really detrimental, by causing lactation to be persevered in, without apparent injury at the time, by females whose bodily vigour is not adequate to sustain it. Such certainly appeared to be the case in the instance just referred to; for the system remained in a very depressed state for some time after the conclusion of the first lactation; and on subsequent occasions it has been found absolutely necessary to discontinue nursing at a very early period of the infant's life, owing to the inadequacy of the milk for its nutrition, and the obvious inability of the mother to bear the drain. Hence it may be affirmed with tolerable certainty, that the first lactation, although not prolonged beyond the usual period, and although apparently well sustained by the mother, was really injurious to her; and the inability to furnish what was required, without the stimulus of alcoholic liquors, was Nature's warning, which ought not to have been disregarded.

202. Considering, then, that lactation (unlike pregnancy) may be put an end to at any period, should it prove injurious to the mother, the writer is disposed to give his full assent to the *dictum* of Dr. Macnish; that "if a woman cannot afford the necessary supply without these indulgences, she should give over the infant to some one who can, and drop nursing altogether." — "The only cases," continues Dr. M., " in which a moderate portion of malt liquor is justifiable, are when the milk is deficient, and the nurse averse or unable to put another in her place. Here, of two evils, we choose the least, and rather give the infant milk of an inferior quality, than endanger its health, by weaning it prematurely, or stinting it of its accustomed nourishment." [1] Now upon this the writer would remark,

---

[1] Anatomy of Drunkenness, p. 301.

that a judicious system of feeding, gradually introduced from a very early period in the life of a child, will generally be preferable to an imperfect supply of poor milk from the mother; [1] and that if the mother be so foolish as to persevere in nursing her infant, when Nature has warned her of her incapacity for doing so, it is the duty of the medical man to set before her, as strongly as possible, the risk—the almost absolute certainty—of future prejudice to herself. The evils which proceed from lactation, protracted beyond the ability of the system to sustain it, may be to a certain degree kept in check by the use of alcoholic stimulants; but the writer is convinced from observation of the above and similar cases, that its manifestation is only postponed. Under no circumstances, therefore, can he consider that the habitual or even occasional use of alcoholic liquors, during lactation, is necessary or beneficial.

Childhood.—It has been maintained by some, that there are certain states of the constitution in *childhood*, in which benefit is derived from the habitual use of small quantities of fermented liquors; and this especially in those who inherit the scrofulous diathesis, [2] and in whom the nutritive functions are altogether imperfectly performed. Experience, it is said, demonstrates the benefit which is derivable from the judicious employment of stimulants, in exciting the digestive and assimilative processes to augmented activity, and in thus improving the general tone of the system. It is not denied that *temporary* benefit may be derived from such a course; but this will be obtained at the risk of prospective evil, extending through the whole of life. For if the habit be begun thus early, it will seldom be found possible to discontinue it; the stomach is rendered dependent upon artificial support; and the improvement which this appears to produce will probably render the parent less anxious to avail himself of *other* means of invigorating the system, and of promoting a more active and complete performance of the nutritive actions, which are more permanent in their character, because they act more *naturally* on the system. Every measure of this kind, therefore,—such as unlimited exposure to fresh air (avoiding damp and cold), plenty of exercise, warm but not too impervious clothing, the copious use of cold water with the addition of salt, sea-bathing, and other adjumenta, [3]—ought to have a complete trial, before recourse be had to the completely artificial support yielded by alcoholic liquors.

204. There cannot be any reasonable doubt that the habitual use of Alcoholic liquors by children in average health, is in every way

---

[1] The author has found in his own experience, that good Cow's milk somewhat diluted with water, and sweetened with a small quantity of sugar (so as to be brought nearly to the composition of Human milk), has answered extremely well even for very young infants.

[2] *Scrofulous diathesis* — constitutional tendency to scrofula.

[3] *Adjumenta* — assisting remedies.

injurious.[1] In no period of life are the nutritive functions more energetically carried on, if the child be only placed in circumstances favourable to health; and at no period of life is there such a disposition to take just that amount of exercise of the nervo-muscular apparatus which is beneficial to the system, without exceeding it. The motives which stimulate the adult to over-exertion in his battle with the world, do not operate upon the child: unless forced by the zeal of injudicious instructors, he will seldom be disposed to carry nis mental exertions beyond the stage at which they may be best intermitted; and whilst naturally prone to muscular exercise, he readily complains of fatigue, and is indisposed to persevere after this warning of the failure of his powers. The chief thing to be watched for and avoided, therefore, is the excess in diet to which children are sometimes prone, more especially if their palates be tempted by articles of which they are fond; and if this be duly restrained, and every *natural* means for the preservation and improvement of health be judiciously and perseveringly employed, it is believed by the writer that more good will in the end be done, than will be accomplished by the assistance of alcoholic liquors. And in support of this belief, he can appeal to the large numbers of families now growing up in this country and in America, in the enjoyment of vigorous health, among whom no alcoholic liquor is ever consumed; and he can point to numerous cases within his personal knowledge, in which the apparent debility of constitution having been such, as in the opinion of some to call for the assistance of fermented liquors, the advice was resisted, and those other means adopted which have been already adverted to, with the effect of rearing to vigour and endurance, children that originally appeared very unlikely to possess either.

205. *Old Age.*—It has been maintained again, by some of those who fully admit the undesirableness of the habitual use of Alcoholic liquors during the vigour of early and middle life, that they are requisite or useful for the support of *old age.* Now upon this point, also, the writer believes that much misconception is preva-

---

[1] In illustration of the injurious effects of the habitual use of fermented liquors upon healthy children, Dr. Macnish (Anatomy of Drunkenness, p. 302,) relates the following experiment made by Dr. Hunter upon two of his children, both of them having been previously unused to wine. To one, a child of five years of age, he gave every day a full glass of sherry; to the other, a child of nearly the same age, he gave an orange. In the course of a week, a very marked difference was perceptible in the pulse, urine, and evacuations from the bowels of the two children. The pulse of the first child was raised, the urine high coloured, and the evacuations destitute of their usual quantity of bile. In the other child, no change what ever was produced. He then reversed the experiment; giving to the first the orange, and to the second the wine, and the results corresponded: the child who had the orange continued well, and the system of the other go straightway into disorder as in the first experiment.

lent, arising out of a disregard to the dictates of Nature on the subject. During the most active period of life, the "waste" of the body is considerable; and the demand for food, and the power of digesting it, are both adequate (in the healthy state) to supply that waste. But with the advance of years, the power of activity diminishes; the body (so to speak) lives much more slowly, as is proved by the lessened exhalation of carbonic acid and the diminished excretion of urea; and the waste being thus lessened, the demand for food, and the power of digesting it, are proportionably diminished. Now this abatement of the appetite and digestive power (like that which is felt by those who go from cold or temperate climates to reside in tropical regions) is a natural warning that a smaller amount of food should be taken in; and if it be so received, and no more nutriment be habitually ingested than the appetite legitimately prompts, the digestive powers will be found as adequate as in a state of greater activity, to provide for the wants of the system. But this abatement is very commonly regarded as an indication of the failure of the powers of the stomach; and recourse is had to alcoholic liquors with the view of re-exciting these. Now although from such a practice, when very moderately resorted to, less prospective evil may be anticipated, as regards merely the effects of the continual ingestion of alcohol upon the *stomach*, than it is liable to occasion when commenced earlier in life; yet it is very much to be deprecated on another account, — namely, that it forces admission into the *system* (so to speak) for a larger amount of alimentary matter than it can appropriate; and as all the organs which are set apart for the elimination of the superfluity (the kidneys, the liver, the skin, and the glandulæ of the intestinal canal,) are less easily stimulated to increased activity in the decline of life than at an earlier period, it follows that habitual excess in diet, even though to no great amount, is yet more likely to be followed by the disorders which it tends to produce. And hence it is, more especially, that we find the lithic acid diathesis [1] so prone to manifest itself in advanced life, and requiring such careful dietetic management for its correction.

206. The author would not take upon him to deny that cases *may* present themselves, in which the habitual use of a small quantity of Alcoholic liquors may be beneficial to persons advanced in life and not suffering under any positive ailment, but experiencing absolute deficiency of digestive power *beyond* that which is in conformity with the general decline of activity; in such cases, the benefit to be expected from their employment is, that the stomach should be assisted in the digestion of the food which the system really requires;

---

[1] *Lithic acid diathesis* — constitutional tendency to the formation of lithic acid, giving rise to stone in the bladder.

and in so far as their use is carried beyond that point, it is hurtful in every way. Such cases may be expected to be rare among those who have habitually observed the laws of health, and who have not prematurely exhausted the powers of their digestive apparatus, by habitual excess in diet or in mental labour, or by the continual use of stimulants. Those, on the other hand, who have adopted the habit, early in life, of relying upon the aid of alcoholic liquors, for the performance of the digestive operation; or who have overtasked their nervous systems, and thus deprived the stomach of the nervous power which it requires; or who have impaired their vigour by breathing a foul atmosphere, by irregularity and insufficiency in regard to the periods of repose, or by various other departures from the ordinances of Nature; are more likely to suffer in advanced life from a loss of digestive power, which no treatment, medical or hygienic, can ever completely repair.

207. But here, as in all other instances, if the prolongation of life and the restoration of vigour be the paramount objects of consideration, recourse should at first be had to all those measures of general Hygiène,[1] which prudential experience would recommend; and the assistance of Alcoholic liquors should be avoided, with a jealous apprehension of their prospective evils, until it shall appear that no other more natural means can bring about the desired result. Those who have been in the habit of employing them during the whole of life, are certainly those who are least likely to feel able to dispense with them in old age; yet experience has demonstrated even here, that where the evil results of their continued use have begun to manifest themselves, decided and permanent benefit has followed their abandonment; and where it was believed by the individual that he could not possibly dispense with their use, the stomach has recovered its healthy tone (especially under the copious external and internal use of cold water, and the influence of an invigorating atmosphere), so as to be able to discharge its duties for the remainder of life with greater ease than it ever previously had done.[2]

208. For the results of experience on this and other points, any statements of which should be based rather on a wide and general survey, than on induction from a comparatively limited number of instances, the author has thought it safest to rely on the assurances of medical practitioners in the New England States; since the entire disuse of fermented liquors has been now practised as a habit for some years, by a large proportion of the population of those States, including those who are most subject to those influences (the

---

[1] *Hygiène* — the art of maintaining and improving health.

[2] For two remarkable cases of this kind, see Appendix C.

14

"wear and tear" of social life) which are usually regarded as most powerfully conspiring to render the assistance of stimulants desirable. — The following statements on this subject have been recently put forth by the Massachusetts Temperance Society, under the sanction of their distinguished President, Dr. Warren.[1]

209. "In regard to the habitual use of wine, it is probable that the change of opinion is greater here than in Europe. A vast number of persons on this side of the Atlantic have wholly abandoned the use of wine, cider, and malt liquors; and many of those who continue to employ them have greatly diminished the quantity. Wine is no longer thought necessary in the convalescent stage of fever. Cider, formerly one of the household provisions of almost every family in the North, is rarely seen; and the very trees which produced it are either cut down for fuel, or converted to the production of fruits for food. The stronger beers are quite disused, except among emigrants; and even the milder are employed only in some very light and unstimulating form to allay thirst, principally in the hot season." "The apprehension that a sudden disuse of fermented liquors might be injurious, has been dispelled by a vast number of cases, in which, after long-continued employment, a sudden and total abandonment has taken place, not only without impairing health and comfort, but with positive improvement in strength, activity, and agreeable sensations. How common is it amongst us to see persons who in former times used wine freely, and who have now given it up, present an appearance of mental and bodily vigour they had not exhibited before. The influence of such a change of habit in the wealthier classes has been great beyond calculation, in leading the mass of society to abandon the use of spirits, and to repeat an experiment already made by those whom they are accustomed to respect and follow. Such having been the consequences of the disuse of wine, how desirable is it that all those who have not abandoned it, who wish well to their fellow-men, and are willing to show that they are capable of making the sacrifice they advise, should submit to a privation which they have sufficient reason to believe will be most salutary to themselves and others."

210. The extent of change of habit, in this respect, among the middle and higher classes of society in Boston, and other great towns of New England, may be judged of from the fact, that many of those public festivities, at which the assistance of alcoholic liquors is considered indispensable in this country, are there conducted without any such artificial excitement. "Of late years" (we quote the same authority) "we have had the gratification of witnessing so many exceptions to the former practice, that it appears

---

[1] Preface to the Reprint of an Essay on the Physiological effects of Alcoholic Drinks, from Dr. Forbes's Review, Boston, N. E. 1848.

very probable that the rule will be reversed, and the exceptions change to the opposite side. The great festivals on the Anniversary of National Independenceare in many places celebrated without other stimulus than that of patriotic feeling. The annual ceremonies of our literary institutions, too often stained by lavish draughts of the juice of the grape, are now purified by the effusions of chastened wit, and elevated by the flights of an unclouded imagination. Most of the Universities, and particularly the oldest, and we may perhaps venture to say the most distinguished, have unshackled themselves from the chains of ancient habit. Under the influence of a master-spirit (President Everett) the great annual festival of Commencement at Cambridge University has been accomplished without the aid of wine; and the oldest of our literary fraternities, the Phi Beta Kappa Society, has enjoyed the excitement of a social meeting without the consequent depression from artificial stimulus. Wine is no longer admitted at the yearly convocation of the Clergy, or the assemblage of the Medical profession of this State. The great association of mechanics of the metropolis hold a brilliant triennial feast, from which every kind of alcoholic, fermented, vinous, and other stimulating liquid is wholly excluded." [1]

211. On the whole, then, the writer thinks that Physiology and Experience alike sanction the conclusion, that although there are states of the stomach, in which the diminished appetite and digestive power prevent the reception of an adequate supply of aliment into the system, and in which the assistance of alcoholic liquors is temporarily beneficial, that assistance is rather a *palliative*, than a *cure* of the condition which calls for it; and, if perseveringly had recourse to, is likely to induce a train of evils of its own: whilst, at the same time, by the apparent support which it gives, and by rendering the system more tolerant of the unfavourable influences from which its depression of power may have proceeded, it renders the individual less disposed to seek, in a change of habit, the remedies which will be really effectual. "Thus," as an American physician has remarked to the writer, "where *you* (the English practitioner) recommend to a man losing his digestive power from the fatigue and confinement of a city life, to take wine, porter, or bitter ale with his dinner, *we*

---

[1] In order to give a more exact idea of the importance of the celebrations alluded to above, we have thought it well to state the number of persons that attended them, as nearly as can be ascertained.

Fourth of July Celebration in Faneuil Hall 1000
Commencement at Cambridge University.......................... 300
Festival of Phi Beta Kappa 150
Festival of the Clergy 200
Festival of the Massachusetts Medical Society............:........ 300
Festival of the Massachusetts Mechanics' Association.......... 600

order him out of town, to get fresh air, and the refreshment of idle ness in the country." No man of observation can doubt which of these two systems is likely to be most beneficial in the long run.

212. But, again, the writer believes that there are exceptional cases, arising chiefly from peculiarity of original constitution, in which the want of digestive power is more completely and permanently supplied by the habitual use of a small quantity of Alcoholic liquors, than it can be by any other means within the power of the individual. It may be impossible to predicate in any individual instance, whether this shall be the case or not; but the results of observation appear sufficient to prove, that it would be erroneous to assert dogmatically that it never can be. Still, the evils resulting from the unnecessary employment of stimulants are so great, that recourse should never be had to them, until every other more natural method of sustaining the vital powers has been tried without success; they should never be employed to replace any hygienic requirement, such as fresh air, mental repose, muscular exercise, &c.; and they should be disused whenever it may appear that the necessity for them no longer exists.

CHAPTER IV.

ᖯ THE EMPLOYMENT OF ALCOHOLIC LIQUORS NECESSARY IN THE PRACTICE OF MEDICINE? IF SO, IN WHAT DISEASES, OR IN WHAT FORMS AND STAGES OF DISEASE, IS THE USE OF THEM NECESSARY OR BENEFICIAL?

213. Those who maintain that Alcoholic liquors are not requisite for the ordinary sustenance of Man, or even that they are likely to be rather prejudicial than otherwise when habitually taken in small quantities, — that, in fact, Alcohol is to almost every one a true *poison*, slower or more rapid in its operation, according to the rate at which it is taken, — may still maintain with perfect consistency, that (like many other poisons) it may be a most valuable *remedy*, when administered with caution and discrimination, in various forms of disease. In replying to the above question, we shall first look at the inferences which we may draw from the physiological action of Alcohol, in regard to the conditions of the system in which it is most likely to be useful.

I. — RECOVERY FROM SHOCK.

213. We have seen that Alcohol, when introduced into the circulation, acts as a stimulant in augmenting the force and rapidity of the heart's contractions, and that it also increases the excitability of the nervous system; we have found, moreover, that it supplies the means of keeping up the Animal Heat, which may be advantageously employed when other means are deficient. Hence we should say that alcoholic liquors may be advantageously employed to assist in rousing the system from the effects of agencies of various kinds, which threaten for a time to produce a fatal depression of the vital powers;—such, for example, as severe injuries that produce a violent *shock*, under the primary effect of which the system appears likely to sink. But great caution must be used in their administration, and they should not be given unless there appears to be a positive necessity for doing so, (*i. e.* unless the patient appears likely to sink without them); for it is as certain that re-actionary excitement will follow a primary depression, as it is that depression will be subsequent upon primary excitement; and if stimulants have been unnecessarily employed, the difficulty of controlling the re-action will be increased. This caution is more especially necessary, where the *brain* is the part to which the injury has occurred, since the special determination of alcohol to this

14 *

organ will increase the violence of the re-action in a most dangerous manner.

214. There is no class of cases, perhaps, in which the good effects of stimulants in maintaining the heart's action, and in keeping up the nervous excitability, are more manifest, than in those severe and extensive burns of the trunk of the body, to which the children of the lower classes are peculiarly liable, from their clothes taking fire through carelessness or negligence. The shock given by this injury to the delicate and impressible system of the child is often rapidly fatal; the heart's action being extremely depressed, the nervous power reduced, and the body gradually cooling, until its temperature falls to a degree incompatible with the maintenance of life. The writer has witnessed many such cases, in which life seemed to be kept in the body by the frequent administration of a spoonful of cordial, but in which death supervened upon a short intermission of the stimulus,—the nurses in Hospitals being generally possessed with the belief that the little patients *must* die, and being too frequently careless in the employment of the only means by which life can be sustained.

II. — TREATMENT OF ACUTE DISEASES.

215. *Resistance to the depressing influence of Morbific Agents.*[1] —In the class of cases to which reference has just been made, the shock is temporary; and if the patient can be kept alive until the system has recovered from its immediate consequences, a great point is gained. There is another class of cases, in which the depression is produced by a morbific agency, and in which it is of equal importance to keep up the vital powers for a time; since, if they can be sustained for a few hours or days, the patient has a fair chance for recovery. Of such we have examples in many forms of Fever; especially in those which run a tolerably determinate course, and which exert their noxious influence rather in producing a general depression of the vital powers, than in occasioning any decided local lesion. No two epidemics of fever are precisely alike; and the treatment which is of service in one may be found injurious in the other, notwithstanding that the general type may be the same. A severe epidemic of typhoid fever, which the writer witnessed in Edinburgh in the years 1836-7, afforded him an opportunity of seeing the decided efficacy of Alcoholic stimulants in one form at least of this fever; the opposite methods of treatment, followed by two physicians whose practice he watched, being attended with such different results, that as the cases were of the same class, and the other conditions identical, there was no other way of accounting for

---

[1] *Morbific agents* — disease-producing agents.

the difference. By neither physician were any active measures taken during the early stages of the fever, for none seemed called for; but in one set of cases, the same expectant practice was continued to the end; whilst in the other, the administration of wine and spirit was commenced, as soon as the weakness of the pulse, and the coldness of the extremities, indicated the incipient failure of the circulating and calorifying powers.[1] The quantity was increased as the necessities of the patient seemed to require; and in one case (that of a woman whose habits had been previously intemperate, and on whom a more potent stimulus was therefore needed to make an impression,) a bottle of sherry with twelve ounces of whiskey was the daily allowance for a week or more,—the patient ultimately recovering. Now the result of this wine-treatment was, that the mortality was *not above a third* of that of the simple expectant treatment; the patients dying under the latter from actual exhaustion and failure of calorifying power, and no local lesion[2] being detectible on post-mortem examination.

216. It is by no means difficult to give a satisfactory *rationale*[3] of this beneficial action. The immediate cause of death in such cases appears to be a failure of the power of the heart, the contractions of which, in the advanced stage of typhus and typhoid fevers, become progressively feebler and more rapid; and it has been noticed by Drs. Stokes and Graves, as the best indication for the use of wine, that the impulse is greatly diminished, and that the first sound becomes very feeble or is entirely extinguished. Now the effect of wine, where it acts beneficially, is to render the heart's action more vigorous and at the same time slower. Again, with this state of the circulation we generally have a low muttering and restless delirium, with an approach to subsultus tendinum;[4] and if the wine acts beneficially, it renders the patient more tranquil and disposes him to sleep. Under the influence of wine, too, in suitable cases, the skin and tongue become moister, and the breathing more deep and slow; but if the wine be acting injuriously, the skin and tongue become drier, and the respiratory movements more hurried. Concurrently with the failure of the heart's action, there seems often to be a deficiency of heat-producing material; all that was previously contained in the body having been burned-off during the earlier period of the fever; and little or none having been taken in from without. Day by day, the fatty matter of the body is used up by the respiratory process; and thus, as in cases of simple starvation, the patient must die of *cold*, unless some means be provided for the

---

[1] *Calorifying powers* — heat-producing powers.
[2] *Lesion* — change of structure or condition produced by disease.
[3] *Rationale* — explanation.
[4] *Subsultus tendinum* — starting or spasmodic twitches of the tendons

sustenance of the heat. In such a condition of the system, no farinaceous or oleaginous matters could be digested or absorbed in sufficient quantity ; whereas alcohol is taken into the current of the circulation by simple endosmose,[1] without any preparation whatever,
and can be immediately applied to the production of heat.

217. Now in the cases in which the Alcohol is thus useful, there
is *an entire absence of stimulating effects.* This is probably due in
part to the fact, that the Alcohol is burned-off nearly as fast as it is
introduced (the general rule in such cases being to give a small quan
tity at a time, but to repeat this frequently) ; but it would also appear
to result in part from this,—that the *stimulating* power of the Alcohol is expended in neutralizing (so to speak) the *depressing* influence
of the fever-poison already in the system, and that it simply tends,
therefore, to restore both the heart and the brain to their condition
of normal activity. Where the habits of the patient have been previously intemperate, the ordinary doses of alcoholic stimulants have
no perceptible effect; and it is necessary to go on increasing them,
until some marked influence is exerted by them, — as in the case
just now cited.

218. It is not only in the idiopathic typhoid and typhus fevers,
that Alcohol thus becomes the most important remedy which the
Physician has at his command; for it is equally so in the typhoid
states of other diseases, especially Erysipelas and the Exanthemata;[2]
and it is in the typhoid form of erysipelas, which so often presents
itself in men of the bad habit of body resulting from habitual intemperance (§ 63), that the largest quantities of alcoholic stimulants
may be given, without any other perceptible effect than that most
beneficial one—the support of the system whilst the disease runs its
course.

219. *Recovery from states of Prostration.*—During the stage of
convalescence from fevers and acute inflammatory diseases, in which
the vital powers have been greatly depressed, it will frequently happen that the use of alcoholic liquors will be decidedly beneficial ;
and this apparently in two ways, — by raising the nervous system
from that low irritative state which is the consequence of depressed
vital power, and by increasing the digestive power of the stomach
and the general nutritive activity of the system, so that the reparative processes take place more rapidly, and the general vigour is
more speedily restored. Every practical man must have perceived,
that the state of debility in which the patient is left after the termination of an acute disease, is extremely different from the state of
exhaustion consequent upon a long-continued course of over-excitement. The former partakes of the nature of *shock;* the vital powers

---

[1] *Endosmose* — the passage of water through organized membranes.
[2] *Exanthemata* — cruptive fevers.

are not so much *exhausted* as *depressed;* and recovery is best promoted by arousing the system, so far as possible, to the due performance of its functions. If alcoholic stimulants are really beneficial under such circumstances, they make their utility apparent in the same way as in the advanced stage of typhoid fever,—that is, by reducing the rapidity of the heart's action at the same time that its strength increases, and by calming the mind instead of exciting it. Dr. A. Combe mentions the case of a delicate lady, who, during recovery from fever, took to the extent of a bottle of Madeira in twenty-four hours, with these obviously beneficial results. It is well known that much depends, in this condition, on procuring as speedy a renewal as possible of the normal actions of nutrition; especially where either the disease, or the treatment it has required, has caused them to be greatly lowered, or almost entirely suspended; for there is great danger lest the convalescent should pass into a cachectic [1] condition, and a foundation be laid for tubercular [2] or other forms of disease dependent upon the imperfect performance of the nutritive processes. Hence, if when these operations are just being renewed, a little increased energy can be artificially imparted to them, we have a better hope of escape from these evil consequences. As a general rule, no alcoholic stimulants should be employed, until after the complete subsidence of the inflammatory processes: but this rule is not invariable; for a state of chronic inflammation is often kept up by the low and imperfect state of the general nutritive operations, and hence, (as Prof. Alison was wont to teach and to practise with great success) however contradictory it may at first appear, we may frequently combine a *general* tonic or somewhat stimulant regimen with *local* depletion or counter-irritation. [3]

220. When Alcoholic stimulants are employed for these purposes, the greatest care and watchfulness should be used in their administration; both to avoid doing positive mischief by an over-dose; and also to avoid bringing the system into a habit of dependence upon them, and thereby predisposing it to the various remoter evils formerly described. There is no doubt that a course of over-indulgence in alcoholic liquors has frequently commenced with the therapeutic [4] use of them; and it is extremely desirable, therefore, that the medical practitioner should enforce the diminution of the dose, and the final discontinuance of the remedy, at the earliest possible period,—substituting, if he should think it necessary, a small quantity of alcohol in some *medicinal* form,—in order

---

[1] *Cachectic* — a bad condition of the body.

[2] *Tubercular* — consumptive; scrofulous.

[3] *Counter-irritation* — irritating substances, as blisters, mustard poultices, &c., applied to the part with the view of counteracting deep-seated or internal inflammation.

[4] *Therapeutic* — medicinal; curative.

that the patient may have as little motive as possible for continuing its use, after the time for their really beneficial action has passed.

221. *Support under Exhausting Drains.*—There is another class of cases, in which the stimulating action of Alcoholic liquors may be occasionally had recourse to with advantage; those, namely, in which there is great drain upon the nutritive material, owing to some disordered action which at the same time lowers the vital powers of the system,—such, especially, as an extensive suppurating[1] surface. Now here the general rule, that the appetite and the digestive power are proportionate to the demand for nutriment in the body does not hold good; since the depressing influence of the disease lowers the functional activity of the digestive apparatus, to such a degree that it cannot supply what is needed; and thus there is a progressive diminution of the nutritive solids of the blood, which still further depresses the vital powers of the system. We should therefore anticipate a beneficial result from such an employment of alcoholic stimulants, as would for a time augment the digestive power of the stomach, and would thus enable it to appropriate and prepare the amount of nutritive matter which the system really needs, whilst at the same time its general powers are sustained under the depressing influence of the disease. Experience shows that such is the case; and that under such circumstances, alcoholic liquors may be beneficially employed, not so much to stimulate the heart, or the nervous system, nor to take the place of solid food; but, by stimulating the stomach, to augment the quantity of solid material which it can advantageously receive.

222. Allusion has already been made to the unfavourable course which febrile and inflammatory diseases are disposed to run in the habitually intemperate; this being chiefly dependent upon the imperfect elaboration[2] of plastic[3] material, which predisposes to suppurative action, or to gangrenous[4] or phagedenic[5] ulceration, and impedes the attempt at regeneration[6] which constitutes a most important part of the sthenic[7] form of inflammation. A similar disposition to the asthenic[8] form of inflammatory disease and its severe consequences, is seen among the habitually ill-fed, ill-lodged, ill-clothed inhabitants of the densest and worst-drained parts of our great towns, many of whom are also intemperate; and in many of these cases, it would seem requisite to support the system by Alco-

---

[1] *Suppurating* — discharging matter.
[2] *Elaboration* — formation; manufacture.
[3] *Plastic* — adhesive; aptness to become organized.
[4] *Gangrenous* — disposed to mortification.
[5] *Phagedenic* — spreading; corroding.
[6] *Regeneration* — formation of new substance.
[7] *Sthenic* — vigorous; active.
[8] *Asthenic* — feeble; accompanied by debility.

holic liquors, even during the acute stage of an inflammatory attack, in order to enable it to resist the depressing influence of the disease, and to bear the requisite treatment. Whatever augments the plasti-'city' of the fibrine, up to a certain point, is likely to be beneficial; and as the great object in such cases is to give the requisite support without stimulus, the use of malt liquors will be indicated. Here, too, we find that experience is in full accordance with the teachings of theory; and that ale and porter are frequently the physician's and surgeon's main-stay under such circumstances. They must, however, be very guardedly employed; and the test of their beneficial influence will be found in the absence of stimulating effects, and in the improvement of the character of the inflammatory process; which will be made known, where there is purulent[2] discharge, by the conversion of a thin, sanious,[3] fœtid[4] pus into that which is expressively designated by the term "laudable,"[5] and by the stoppage of an extending gangrene or phagedæna.

223 The foregoing are the principal forms of acute disease, in which recourse may be advantageously had to Alcoholic liquors; but the writer would remark that whilst general principles may be thus laid down, their application to each individual case must be left to the discrimination and tact of the practitioner, since no two cases are alike in all their conditions; and it will frequently happen that even the most experienced physician and surgeon will find it necessary to be rather guided by the result of trials cautiously made, than by any rules whatever. In cases of fever, it may be especially noticed that the *instinct* of the patient, shown by his desire for wine, or his disposition to reject it, will generally prove a most valuable guide, even when his *intelligence* is prostrated.

224. *Forms of Alcoholic Liquors most desirable.*—The different forms of alcoholic liquors must not be used indiscriminately in these varying conditions, for their operation upon the system differs considerably, and there are certain conditions of the body to which each is especially appropriate. Thus, distilled spirit is the most rapid and powerful in its action upon the heart and nervous system; and hence it is the most potent form of alcoholic liquor, in those states of alarming depression from which we desire to arouse the patient as rapidly as possible. We find, too, that it is frequently requisite to administer spirits to patients who have been in the habit of free or excessive indulgence in alcoholic liquors, under circumstances in which wine would otherwise be preferable, e g. in Fever; the milder stimulus, in such cases, not producing the effect

---

[1] *Plasticity* — adhesiveness; aptness to become organized.
[2] *Purulent discharge* — discharge of matter.
[3] *Sanious* — bloody.
[4] *Fœtid* — having an offensive smell.
[5] *Laudable* — healthy.

we desire. Where, however, we desire to give more continued support, with less of stimulation, it is not usually desirable to administer distilled spirit, and wine will be found the preferable form ; this is especially the case in the advanced stage of fever, and in convalescence from acute diseases. On the other hand, where we desire to give still greater support with as little stimulation as possible, as in the class of cases last referred to, malt liquor may be more advantageously employed ; as the alcohol, probably from its peculiar state of admixture, is less disposed to exert its remote effects, and the nutritive matter with which it is combined is in itself beneficial ; whilst the bitter and somewhat calmative properties of the hop aid in producing the desired effect upon the stomach.

III. — TREATMENT OF CHRONIC DISEASES.

225. Of the use of Alcoholic liquors in the treatment of *chronic* [1] diseases, however, it becomes us to speak with much greater caution ; the condition of the system under the *depressing* influence of "shock" or of poisonous agents, being very different from that which results from the *exhaustion* of its powers through chronic diseases, although debility is a characteristic of both. The writer's idea of the difference between the two states, and of the relations of each to alcoholic stimulants, may perhaps be best explained by a simple illustration. When a vigorous man is prostrated by a violent blow, he speedily rallies from it, and is all the better for the aid of a helping hand in getting on his legs again. But if the same man expend his powers in a prolonged pugilistic encounter, although he may not have received any one severe injury, he becomes at last so exhausted that no helping hand can avail him anything, and he sinks beneath the slightest force put forth by his opponent, — nothing but time and rest being then effectual for his restoration. In general, then, it is believed by the writer that little permanent good can be expected from the use of alcoholic stimulants in chronic diseases, so far, at least, as regards their stimulant operation upon the heart and nervous system ; and what benefit they are capable of conferring, will be obtained by their improvement of the digestive power, and of the processes of primary assimilation.[2] But it is very doubtful whether the temporary improvement which can sometimes be thus obtained, is not really fallacious, — like that which we see in the burning of a lamp, just after the raising of the wick, when there is a deficiency of oil ; — since it is procured, not by the re-animation of power which exists

---

[1] *Chronic* — slow ; long-continued.

[2] *Assimilation* — conversion of food into living matter corresponding with the organs of the body.

in the body but has previously lain dormant, but by the more rapid consumption of the small stock of power left. And the question of their benefit or injury will often depend upon whether, by this more rapid consumption, new vigour can be infused into the system, by the introduction of new material.

226. The use of Alcoholic stimulants in such cases has been hitherto so customary with medical men, that it may seem to be running in the face of the established results of experience, to deny or even to doubt their efficacy. But we have seen reason, within a recent period, to deny or doubt the efficacy of many systems of treatment of chronic diseases, which long-continued experience appeared to have sanctioned, and to believe that the *vis medicatrix*[1] of the system is often itself the great restorer, when time is given for its operation, and other circumstances concur to favour it. And it is especially important, in judging of the action of all remedies which must be persevered in for some time in order to produce any decided result, to take their remote consequences fully into account, and to consider how far these are, or are not, favourable to our object. Now the writer has endeavoured to show that the remote consequences of the continued use of alcoholic liquors, even in small quantity, are all of them so unfavourable to health, that if the immediate invigoration of the digestive power and of the assimilative processes, which seems to be their *only* beneficial effect under such circumstances, can be obtained in some other way, it will be most desirable to avoid their use.

227. This will be more particularly the case, when the causes of the disordered state have been such as to exhaust the vital energy of the Stomach itself, — such as long-continued excess in diet, and habitual indulgence in a moderate allowance of fermented liquors, especially when accompanied by exhaustion of the nervous power by over-exertion or anxiety of mind. It is quite absurd to expect that any change or variety of direct stimulation can re-invigorate the digestive apparatus under such circumstances. We may keep our patient in town at his usual occupations, practise all kinds of experiments upon his stomach, recommend fat bacon or lean chops, prescribe blue-pill and senna-draught, or quinine and calumbo, and ring the changes upon all the wines, spirits, and malt-liquors which the cellar can furnish, without effecting any permanent benefit. Whereas, if he can be induced to give himself a complete holiday; to betake himself to some agreeable spot, where there is sufficient to interest, but nothing to excite; to inhale the fresh and invigorating breezes of a mountainous country, in place of the close and deteriorated

---

[1] *Vis medicatrix* — the inherent power of the animal system to repair any injury it may receive, or to remove the diseases with which it may be attacked.

15

atmosphere of a town; to promote the copious action of his skin by exercise, sweating, and free ablution; to wash out his inside and increase the tonic power of his stomach with occasional (but not excessive) draughts of cold water; and to trust to the natural call of appetite alone, in preference to artificial provocatives;—we shall be giving him the best possible chance of permanent restoration to health.

228. There is perhaps no class of cases in which the benefits of the Hydropathic treatment are so strikingly displayed, especially when it is carried on in a spot where all other aids concur to make it most effectual; and reasoning from analogy, the writer is strongly inclined to believe that it would be of similar efficacy in re-invigorating the system exhausted by other forms of chronic disease, and would in most cases be preferable to any form of alcoholic stimulants for procuring an increase of digestive and assimilative power. So far as the writer is acquainted with the results of comparative experience, they are decidedly in favour of the Hydropathic treatment, moderately and judiciously applied, especially in cases of chronic Gout and Rheumatism; but he would not be dogmatic enough to assert that there are not individual instances, in which (as in the class formerly adverted to, § 194), the long-continued or even the habitual use of alcoholic liquors, will promote the recovery from chronic diseases, by their influence on the digestive and assimilative apparatus. He does not see the possibility, however, of laying down any general rules, by which such cases can be distinguished; and it will be only from the results of an extended experience of the comparative advantages and disadvantages of different modes of treatment, and of the immediate and remote consequences of the employment of alcoholic stimulants, as compared with those of the abstinent system, that any really valuable inferences can be drawn. Until these shall have been obtained, he believes that abstinence will in most cases be the safer plan; except where the prostration of the vital powers has proceeded to such an extent, as to require temporary stimulation for the performance of any of the nutritive and regenerative operations.

229. It may not be amiss to remark, in conclusion, that it is through the medium of the *Water* contained in the Animal Body, that all its vital functions are carried on. No other liquid than Water can act as the solvent for the various articles of food which are taken into the Stomach. It is Water alone which forms all the fluid portion of the blood, and thus serves to convey the nutritive material through the minutest capillary pores into the substance of the solid tissues. It is Water which, when mingled in various proportions with the solid components of the various textures, gives to

them the consistence which they severally require. And it is Water which takes up the products of their decay, and conveys them, by a most complicated and wonderful system of sewerage, altogether out of the system. It would seem most improbable, then, that the habitual admixture of any other fluid, — especially of one which, like Alcohol, possesses so marked a physical, chemical, and vital influence upon the other components of the Animal body,—can be otherwise than injurious in the great majority of cases; and where a benefit *is* derivable from it, this will depend upon the fact that the abnormal condition of the system renders some one or more of the special actions of alcohol *remedial* instead of *noxious,* so that the balance becomes on the whole in favour of its use.

APPENDIX A, p. 66.

~~~~~~~~~~~~

Some very remarkable details regarding the condition of the ballasters and Coal-whippers employed on the Thames, have lately been ascertained by the inquiries of the "Commissioner" employed by the *Morning Chronicle*, and have been made public in the columns of that newspaper. The drinking habits of these men have been in part engendered by the system under which they have worked; which has involved, as the necessary condition of their employment, the expenditure of a large part of their earnings at the public-house. This system was done away with, a few years since, as regards the Coal-whippers; but it still remains in force with respect to the Ballasters. Several of the former class are now Total Abstainers; while others who adopted the Abstinence system for a time have returned to their former habits. The inquiries of Mr. Mayhew, the "Commissioner," were specially directed to the experience of both these classes; and we shall first quote the statement of one of the latter, by way of showing the extreme severity of the labour undergone by these men, and the circumstances under which the assistance of Alcoholic liquors is sought by them : — "I was a strict teetotaler for many years, and I wish I could be so now. All that time I was a coal-whipper, at the heaviest work, and I have made one of a gang that has done as much as 180 tons in one day. I drank no fermented liquors the whole of the time. I had only ginger-beer and milk, and that cost me 1s. 6d. It was in the summer time. I didn't 'buff it' that day; that is, I didn't take my shirt off. Did this work at Regent's Canal, and there was a little milk-shop close on shore, and I used to run there when I was dry. I had about two quarts of milk and five bottles of ginger-beer, or about three quarts of fluid altogether. I found that amount of drink necessary. I perspired very violently — my shirt was wet through, and my flannels wringing wet with the perspiration over the work. The rule among us is that we do twenty-eight tons on deck, and twenty-eight tons filling in the ship's hold. We go on in that way throughout the day, spelling at every 28 tons. The perspiration in the summer streams down our foreheads so rapidly, that it will often get into our eyes before we have time to wipe it off This makes the eyes very sore. At night when we get home we cannot bear to sit with a candle. The perspiration is of a very briny nature, for I often taste it as it runs down my lips. We are often so heated over our work that the perspiration runs into the shoes; and often, from the dust and heat, jumping up and down, and the feet being galled with the small dust, I have had my shoes full of blood. The thirst produced by our work is very excessive. It is completely as if you had a fever upon you. The dust gets into the throat, and very nearly suffocates you. You can scrape the coal-dust off the tongue with the teeth; and do what you will, it is impossible to get the least spittle into the mouth. I have known the coal-dust to be that thick in a ship's hold, that I have been unable to see my mate, though he was only two feet from me. Your legs totter under you. Both before and after I was a teetotaler, I was one of the strongest men in the business. I was able to carry seven hundred weight on my back for fifty yards, and I could lift nine half-hundreds with my right arm. After finishing my day's work I was like a child with weakness."

To the foregoing account, the following may be added from another witness by way of finish to the picture :—

"Then there's the coals on your back to be carried up a nasty ladder or some such contrivance, perhaps twenty feet — and a sack full of coals weighs 2 cwt. and a stone at least; the sack itself's heavy and thick. Isn't that a strain on a man? No horse could stand it long. Then when you get fairly out of the ship you go along planks to the waggon, and must look sharp, 'specially in slippery or wet weather, or you'll topple over, and there's the hospital or work-house for you. Last week we carried along planks sixty feet at least. There's nothing extra allowed for distance, but there ought to be. I've sweat to that degree in summer that I've been tempted to jump into the Thames just to cool myself. The sweat's run into my boots, and I've felt it running down me for hours as I had to trudge along. It makes men bleed at the nose and mouth, this work does. Sometimes we put a bit of coal in our mouths to prevent us biting our tongues."

Now it cannot be questioned that such labour is greater than any man should be called on to perform; and that even if it should be proved that assistance is derived in its execution from the use of Alcoholic liquors, the fact would not be in the least degree in their favour. For we might fairly anticipate that under this artificial stimulation, more work being got out of the frame than it is naturally capable of discharging, its powers would be exhausted at an earlier period than that to which they would be preserved under a system of less excessive labour, performed without artificial support. And that such is the case, is abundantly proved by the fact, that such of these men as survive the attacks of acute disease, or

15*

are not the subjects of accident, become prematurely old ; and tha, among the whole class, there are few who have passed the age of fifty years. The amount of Alcoholic liquor habitually consumed by them may be judged of from the following statements made to Mr. Mayhew, by two men who have remained firm to the Total Abstinence principle. —" Before I was a teetotaler I principally drank ale. I judged that the more I gave for my drink the better it was. Upon an average I used to drink from three to four pints of ale per day. I used to drink a good drop of gin too. The coal-porters are very partial to dog's-nose — that is, half-a-pint of ale with a pennyworth of gin in it; and when they have got the money, they go up to what they term the ' lucky-shop' for it. The coal-porters take this every morning through the week, when they can afford it. After my work I used to drink more than when I was at it. I used to sit as long as the house would let me have any. Upon an average, I should say, I used to take three or four pints more of an evening; so that altogether I think I may fairly say I drank my four pots of ale regularly every day, and about half-a-pint of dog's-nose. I reckon my drink used to cost me 13s. a week when I was in work. At times I was a noisy drunken gentleman then."

Another coal-porter, who has been a teetotaler ten years on the 25th of last August, told Mr. M. that before he took the pledge he used to drink a great deal after he had done his work, but while he was at his work he could not stand it. "I don't think I used to drink more than three pints and-a-half and a pennyworth of gin in the day time," said this man. "Of an evening I used to stop at the public-house generally till I was drunk, and unfit to work in the morning. I will vouch for it I used to take about three pots a day after I had done work. My reckoning used to come to about 1s. 8d. per day, or including Sundays, about 10s. 6d. per week. At that time I could average all the year round 30s. a week, and I used to drink away ten of it regularly ! I did indeed, sir, more to my shame."

It seems a legitimate inference, from the early decay of the physical powers of these men, that no real support is given them by Alcoholic liquors, in the performance of their arduous labour; and it is a remarkable point in the statements just quoted, that both agree in the assertion, that the principal part of the liquor consumed is taken in the evening, after the day's toil is over, as they "could not stand it" whilst at work. Thus it appears that the amount which can be effectively employed as a stimulus to nervo-muscular exertion is really small; and it is further evident that there is an entire absence of proof that anything is in the end *gained* by their use ;—a conclusion which is in perfect harmony with the statements made in the Essay (§§ 85–102), as to the incapacity of Alcoholic liquors for maintaining the physical powers of the human system.

That some of those who have tried the Total Abstinence system have gone back to their former habits, from a feeling of the necessity of support, is capable of being accounted for, not merely by the excessive amount of labour they are called on to perform, but also by the want of adequate sustenance from solid food. A due allowance of this is unquestionably essential to the maintenance of the strength; and it appears from the testimony of other individuals similarly employed (given in the next Appendix), that where this course has been followed, the labour has been performed with more ease, and that the power of endurance has been increased.

APPENDIX B, p. 87.

THE recent inquiries of the "Commissioner" of the *Morning Chronicle*, not only reveal the circumstances under which a vast amount of excessive drinking takes place among men engaged in laborious employments, but also confirm all that is stated in the text as to the possibility of performing the severest labour without such assistance, provided due support be obtained from solid food. The following is the statement made to Mr. Mayhew by a Coal-whipper, who had been a teetotaler of eight years' standing: — " It's food only that can give real strength to the frame. I have done more work since I have been a teetotaler in my eight years, than I did in ten or twelve years before. I have felt stronger. I don't say that I do my work better; but this I will say, without fear of successful contradiction, that I do my work with more ease to myself, and with more satisfaction to my employer, since I have given over intoxicating drinks. I scarcely know what thirst is. Before I took the pledge I was always dry; and the mere shadow of the pot-boy was quite sufficient to convince me that I wanted something. I certainly havn't felt weaker since I have left off malt liquor. I have eaten more and drank less. · I live as well now as any of the publicans do—and who has a better right to do so than the man who works? I have backed as many as sixty tons in a day since I took the pledge, and have done it without any intoxicating drink with perfect ease to myself, and walked five miles to a temperance meeting afterwards. But before I became a teetotaler, after the same amount of work I should scarcely have been able to crawl home. I should have been certain to have lost the next day's work at least; but now I can back that quantity of coals week after week without losing a day I've got a family of six children under twelve years of age. My wife's a teetotaler, and has suckled four children upon the principle of total abstinence. Teetotalism has made my home quite happy, and what I get goes twice as far. Where I work now, four of us out of five are teetotalers. I am quite satisfied that the heaviest work that a man can possibly do, may be done without a drop of fermented liquor. I say so from my own experience. All kinds of intoxicating drinks is quite a delusion. We teetotalers can do the work better, that is, with more ease to ourselves, than the drinkers can. Many teetotalers have backed coals out of the hold, and I have heard them say over and over again that they did their work with more comfort and ease than they did when they drank intoxicating drink. Coal-backing from the ship's hold is the hardest work that it is possible for a man to do. Going up a ladder sixteen feet high, with 238 lbs. weight upon a man's back, is sufficient to - kill any one; indeed, it does kill the men in a few years — they're soon old men at that work."

It appears from the statement of this man, — 1st, That he could do more work in the time, on the total abstinence system, than on his previous system;—2nd, That he could do it with more ease and comfort to himself; — 3d, That at the end of his day's labour he was not too much fatigued to prevent him from voluntarily walking five miles to a temperance meeting; whilst (like the first witness cited in the preceding Appendix, who spoke of himself as "like a child with weakness" when his toil was over,) he was previously scarce able to crawl home.

APPENDIX C, p. 157.

On the effect of Water-Drinking in the cure of Gout; by JOHN BOSTOCK, M.D., F.R.S. — *Communicated to the Medico-Chirurgical Society.*

"THE case to which I propose to direct the attention of the Society, is that of a gentleman seventy years of age, who had been, from a very early period of his life, subject to very frequent attacks of gout, the predisposition to which complaint is inherited from his parents. Connected with this, he has been a constant sufferer from stomach-affections of various kinds; acidity, flatulence, heart-burn, irregularity of the bowels, and in short, from every one of the affections which are enumerated in Cullen's well-known definition of dyspepsia. His mode of life was regular and moderately active, and his diet what might be styled temperate, although not abstemious. He had, indeed, been advised by his medical friends to take wine in moderate quantity; he had occasionally employed ale, porter, and brandy-and-water, but never in what could be considered an excessive quantity. In this way he had passed about forty years, seldom actually confined by indisposition, but almost always subject to a succession of ailments, which rendered it necessary to have recourse to medicines of various kinds, and, more especially, to alkalies, which were taken in large quantity, and, as the symptoms appeared to indicate, to purgatives or to sedatives, and to a variety of tonics and stimulants. During this period, the renal secretion was seldom in what could be considered a perfectly healthy state; it was sometimes loaded with deposits, and of high specific gravity; sometimes of low specific gravity, limpid and aqueous; sometimes very copious, at other times scanty; while its chemical constitution was most variable both as to the nature and the proportion of its saline contents.

" About four years ago, in consequence of the accession of certain alarming symptoms of a new description, which were supposed to require the antiphlogistic treatment, the patient was ordered by his medical attendants to reduce his system of diet, and, more especially, to abstain entirely from fermented liquors or distilled spirits of any description. By this restriction, and by other appropriate remedies, the threatened disease was averted. And besides this fortunate result, the patient found his general state of health and feelings so much improved by the change of diet, that the abstinence from all kinds of liquors has been strictly adhered to up to the present period. The effect has been *that he has lost all the dyspeptic symptoms to which he had been subject for upwards of forty years ;* and, what I am more particularly desirous of pointing out to the Society, *the renal secretion has been now, for a long period, in a perfectly natural state :* it is nearly uniform in its specific gravity, and is totally free from all the morbid deposits, which were before seldom absent from it. And there is a circumstance connected with it, which I conceive to be particularly deserving of attention ; that although of an average specific gravity, and containing the proper proportion of urea and saline ingredients, it is uniformly increased in quantity, so that there has been now, for several months, considerably more of these substances discharged from the system than was formerly the case. It would appear, therefore, that the abstraction of alcohol has produced a more healthy state of the digestive and secreting functions ; so that the functions of the kidney are more actively and effectively performed."—*Medical Gazette,* Feb. 23*rd,* 1844.

To this interesting case, which is understood to be that of Dr. Bostock himself, may be added the following, from the *Bristol Temperance Herald.*

" Rebecca Griffiths, the individual referred to, resided in this city (Bristol) the larger portion of her long life, and until her *eighty-ninth* year had daily taken as a beverage some kind of intoxicating drink. Beer, and occasionally gin-and-water, had been commonly used; but for a few years before practising total abstinence, she took daily a small portion of the best Madeira wine — having, perhaps, both as regarded the quantity and quality of the liquor, every advantage that any one could possess in using a stimulating drink. This practice she relinquished all at once in the eighty-ninth year of her age. For a time it was feared her health would suffer, but it was soon manifest that those fears were groundless; her appetite improved with the change of diet, and occasional interruptions by a disordered stomach were much less frequent; she would at times observe, that she could eat, drink, and sleep, as well as at almost any period of her life; nor did her spirits appear to suffer even temporarily. For nearly fifteen years she had been tried with a sore in one of her legs, which was troublesome, and at times appeared to be dangerous; after practising teetotalism for about a year and a half, this sore began to diminish, and was soon perfectly healed. At the expiration of two years she had a rather violent attack of influenza, which brought her so low that her medical attendant recommended wine, to which she had recourse for about six months, *when the wound in her leg again opened,* and became troublesome; the wine was consequently discontinued, and after the lapse of a few months the *sore again healed up!* Her health also improved yet more decidedly than after her first trial of total abstinence, and she continued until within a few days of her decease (which took place in the spring of 1843) in the enjoyment of excellent health and spirits, and the full possession of nearly all her faculties, although 93 years old.